Lecture Notes in Biomathematics

Managing Editor: S. Levin

11

Mathematical Models in Medicine

Workshop, Mainz, March 1976

Edited by J. Berger, W. Bühler, R. Repges, and P. Tautu

Springer-Verlag
Berlin · Heidelberg · New York 1976

Editors
Jürgen Berger
Fakultät für Klinische Medizin Mannheim
der Universität Heidelberg
Theodor-Kutzer Ufer
6200 Mannheim/BRD

Wolfgang J. Bühler
Fachbereich Mathematik
Johannes Gutenberg Universität
Saarstraße 21
6500 Mainz/BRD

Rudolf Repges
Abt. Medizinische Statistik
und Dokumentation
Theaterstraße 13
5100 Aachen

Petre Tautu
Institut für Dokumentation
Information und Statistik
Deutsches Krebsforschungszentrum
Im Neuenheimer Feld 280
6900 Heidelberg 1

AMS Subject Classifications (1970): 92A05, 92A15, 76Z05, 60J20, 60J85

ISBN 3-540-07802-9 Springer-Verlag Berlin Heidelberg New York
ISBN 0-387-07802-9 Springer-Verlag New York Heidelberg Berlin

Printed in Germany
Printing and binding: Beltz Offsetdruck, Hemsbach/Bergstr.

PREFACE

On March 8/9, 1976 a workshop on "Mathematical Models in Medicine" was held at Mainz (German Federal Republic) by the group of "Mathematical Models" of the Deutsche Gesellschaft für Medizinische Dokumentation, Informatik und Statistik. Purpose of this conference was to bring together experts from the GFR and neighbouring countries working in this field to evaluate possibilities and limits of this area of research in discussions with interested participants. This issue of Lecture Notes contains the invited contributions as well as the relevant remarks made by the discussants. Corresponding to the aims of the workshop the contributors had been encouraged to demonstrate their mathematical models in the light of actual applied examples.

It had been our intention to restrict attention to a small number of specific areas in order to achieve a concentrated in depth treatment in these restricted areas. The areas chosen contain two - Epidemiology and Cell Models - which in the organisers feeling are not yet as well established in Continental Europe and one - Pharmacokinetics - with a more direct appeal to applied workers. While in the Epidemiology of infectious and parasitic diseases today strategies of control and eradication are gaining importance, the cell models are concerned with explaining the modes of genesis of cancerous growth and the kinetics and interactions within multi-cell structures. The approach in Pharmacokinetics is comparatively more descriptive: the interest here lies with the local concentrations of chemicals in an organism without the declared intention of explaining the underlying regulatory mechanisms. Here more and more the modern computational facilities become an indispensable tool in achieving a realistic description of the net structure of the mechanisms determining changes in concentration.

Mathematical models will only appeal to and be useful for the applied researcher if they are conceived on the basis of real experimental data. A mathematical model successful in describing results from one experiment may also sensibly be used as a filter in the determination future strategies with rescept to the planning of more experiments to increase our knowledge or in the planning of optimal allocation of restricted funds for the control of some disease.

Thanks are due the Merrell Pharma der Richardson Merrell GmbH Groß-Gerau and the Deutsche Gesellschaft für Medizinische Dokumentation, Informatik und Statistik for their generous financial support without which the workshop would never have come true.

J. Berger W. Bühler R. Repges P. Tautu

List of Authors and Organizers

J. Berger	Institut für Medizinische Statistik und Dokumentation Universität Mainz Langenbeckstraße 1 6500 Mainz
W. Bühler	Fachbereich Mathematik Universität Mainz Saarstraße 6500 Mainz
R.M. Anderson	Department of Zoology University of London, King's College Strand, London WC2R 2LS, U.K.
K. Dietz	Statistican, Health Statistical Methodology World Health Organization 1211 Geneva 27 - Switzerland
U. Feldmann	Abteilung für Biometrie Medizinische Hochschule Hannover Karl Wiechert Allee 9 3000 Hannover-Kleefeld
A. Fortmann	Gaußstraße 3140 Lüneburg
W. Müller-Schauenburg	Medizinisches Strahleninstitut Abt. Nuklearmedizin Universität Tübingen Röntgenweg 11 7400 Tübingen
M. Nagel	Institut für Mathematische Maschinen- und Datenverarbeitung II Universität Erlangen-Nürnberg Egerlandstraße 13 8520 Erlangen
R. Repges	Abt. Medizinische Statistik und Dokumentation Rheinisch-Westfälische Technische Hochschule Aachen Theaterstraße 13 5100 Aachen

O. Richter — Institut für Medizinische Statistik
Universität Düsseldorf
Moorenstraße 5
4000 Düsseldorf 1

W. Rittgen — Institut für Dokumentation, Information und Statistik
Deutsches Krebsforschungszentrum
Im Neuenheimer Feld 280
6900 Heidelberg 1

B. Schneider — Abt. für Biometrie
Medizinische Hochschule Hannover
Karl Wiechert Allee 9
3000 Hannover-Kleefeld

K. Schürger — Institut für Dokumentation, Information und Statistik
Deutsches Krebsforschungszentrum
Im Neuenheimer Feld 280
6900 Heidelberg 1

G. Segre — Instituto di Farmacologia
Universita di Siena
Via Banchi di Sotto 55
53100 Siena / Italy

P. Tautu — Institut für Dokumentation, Information und Statistik
Deutsches Krebsforschungszentrum
Im Neuenheimer Feld 280
6900 Heidelberg 1

H.E. Wichmann — Medizinische Klinik
Universität Köln
Josef Stelzmann-Straße 9
5000 Köln 41

List of Participants

(Workshop "Mathematische Modelle in der Medizin" 8 - 9 March 1976)

Dr. Hanns Ackermann
Klinikum der Univ. Frankfurt
Abt. f. Biomathematik
Theodor-Stein-Kai 7
6000 Frankfurt

Dr. R.M. Anderson
Univ. of London
King's College
Dept. of Zoology
Strand London WC 2RLS
England

Dipl.-Math. G. Bach
Abt. Med. Stat. u. Dok. der
Med. Fakultät an der Rheinisch-
Westf. Techn. Hochschule Aachen
Theaterstr. 13
5100 Aachen

Bernd Bausch
Herzinfarktzentrum Heidelberg
Bergheimer Str. 58
6900 Heidelberg

Prof. Dr. J. Berger
Inst. f. Med. Stat u. Dok.
der Univ. Mainz
Langenbeckstr. 1
6500 Mainz

Dr. Bloedhorn
Inst.f. Med. Dok. u. Stat. der
Univ. Köln
Joseph-Stelzmann-Str. 9
5000 Köln 41

Dr. Winfried Blumberger
Troponwerke
Berliner Str. 220
5000 Köln 80

Dr. H. H. Bock
Techn. Universität Hannover
Lehrstuhl f. Mathemat. Stochastik
Schneiderberg 50
3000 Hannover

Ulrich Buschsieweke
Inst. f. Med. Dok. u. Stat.
der Universität Köln
Joseph-Stelzmann-Str. 9
5000 Köln 41

Prof. Dr. W. Bühler
Fachbereich Mathematik der
Universität Mainz
Saarstr.
6500 Mainz

Heinz L. Christl
DKD Wiesbaden
Aukammallee 33
6200 Wiesbaden

Guntram Deichsel
Inst. f. Med. Biometrie
Äulestr. 2
7400 Tübingen-Lustnau

Dr. K. Dietz
World Health Organization HSM
CH 1211 Geneva 27
Schweiz

Dr. Paul Doucet
Biologisches Laboratorium der
Vrije Universiteit
De Boelelaan 1087
Amsterdam

Dr. Manfred Ehrhardt
Med. Klinik u. Poliklinik
(Innere Medizin II)
der Universität des Saarl.
6650 Homburg

Priv.-Doz. Dr. Feldmann
Dept. Biometrie d. Med. Hochschule Hannover
3000 Hannover

Dr. Heinz Fink
Bayer-Elberfeld
5600 Wuppertal

Dr. H. Flühler
c/o CIBA-GIGY AG
WRZ R-1007.2.17
CH-4002 Basel
Schweiz

Dr. A. Fortmann
Gaußstr. 2
3140 Lüneburg

Dr. Rainer Frentzel-Beyme
DKFZ Heidelberg
Im Neuenheimer Feld 280
6900 Heidelberg

Dipl.-Math. Stefan Gräber
Rechenzentrum der Univ.
d. Saarlandes
Nebenstelle Homburg
6650 Homburg

Dr. Winfried Gunselmann
Inst. f. Med. Stat. u. Dok.
Waldstr. 6
8520 Erlangen

Dr. Franz-Josef Hehl
Uni-Klinik Heidelberg
Thibautstr. 3
6900 Heidelberg

Prof. Dr. Herbert Immich
Inst. f. Med. Dok. u. Stat.
Im Neuenheimer Feld 325
6900 Heidelberg

Prof. Dr. H.J. Jesdinsky
Inst. f. Med. Stat. u. Biomath.
Universitätsstr. 1
4000 Düsseldorf 1

Ernst Jurgovsky
Fa. E. Merck
Frankfurter Str. 250
6100 Darmstadt 2

Dr. Friedhelm Knappen
C. H. Boehringer Sohn
Binger Str.
6507 Ingelheim/Rhein

Dr. Jan Peter Knaus
Inst. f. Med. Stat. u. Dok.
der Universität Mainz
Langenbeckstr. 1
6500 Mainz

Dr. Winfried Koch
BASF
6700 Ludwigshafen

Horst Leicher
Fa. Hoechst AG
6000 Frankfurt-Hoechst 80

Dr. med. Rudolf Leutner
Stat. Bundesamt
Günter-Stresemann-Ring
6200 Wiesbaden

Dr. R. Lorenz
Bundesforschungsanstalt für
Viruskrankheiten der Tiere
Paul-Ehrlich-Str. 28 - Postf.1149
7400 Tübingen

Dr. Tom Louton
Inst. f. Med. Stat. u. Dok.
der Med. Akademie Lübeck
Ratzeburger Allee 160
2400 Lübeck

Dipl.-Phys. Rudolf Mader
Abt. f. Med. Dok. u. Datenverarbeitung der Universität
Calwerstr. 7
7400 Tübingen

Cand. med. A. Martini
Pharmakolog. Institut
der Universität Mainz
Langenbeckstr. 1
6500 Mainz

Dr. Gerald Morawe
Abt. Biomathematik
Klinikum Univ. Frankfurt
Theodor-Stern-Kai 7
6000 Frankfurt

Wolfgang Morgenstern
Med. Sozialmedizin Heidelberg
Thibautstr. 3
6900 Heidelberg

Dr. Hermann Moser
c/o CIBA-GEIGY AG
R-1004.2.33
CH Basel / Schweiz

Dr.Dr. W. Müller-Schauenburg
Med. Strahleninstitut der
Universität Tübingen
Abt. Nuklearmedizin
Röntgenweg 11
7400 Tübingen

Dr. Manfred Nagl
Inst. f. Mathem. Maschinen- und
Datenverarbeitung II der
Univ. Erlangen-Nürnberg
Egerlandstr. 13
8520 Erlangen

Dr. Albrecht Neiss
Inst. f. Med. Stat.
Sternwartstr. 2/II
8000 München 80

Dr. Jaak Nijssen
Chemie Grünenthal
Zweifaller Str.
5190 Stolberg

Dipl.-Math. Klaus Pahnke
Fa. E. Merck
Frankfurter Str. 250
6100 Darmstadt 2

Dr. Ulrich Pallaske
Fa. Bayer AG
5090 Leverkusen

Dipl.-Math. W. Pietzsch
Inst. f. Strahlenhygiene
d. Bundesgesundheitsamtes
Corrensplatz 1
1000 Berlin 33

Dipl.-Math. N. Pyhel
Abt. Med. Stat. u. Dok. der
Med. Fakultät an der Rheinisch-
Westf. Techn. Hochschule Aachen
Theaterstr. 13
5100 Aachen

Werner Rehn
Rathausstr. 12
6500 Meinz Bretzenheim

Dr. Hans-Jürgen Reinhard
Inst. f. Biometrie u. Stat.
Bischofsholer Damm 15
3000 Hannover

Prof. Dr. R. Repges
Abt. Med. Stat u. Dok. der
RWTH Aachen
Theaterstr. 13
5100 Aachen

Dr. Claus Rerup
Pharmakologische Institutionen
Sölvegartens 10
22361 Lund/Schweden

Dr. O. Richter
Inst. f. Med. Stat.u.
Biomathematik
Universitätsstr. 1
4000 Düsseldorf

Dr. W.P. Rittgen
DKFZ
Inst. f. Dok., Inform. u. Stat.
Im Neuenheimer Feld 280
6900 Heidelberg 1

Dr. Klaus Rockenfeller
Max-Planck-Institut
Parkstr. 1
6350 Bad Nauheim

Dr. Anton Safer
Fa. Knoll AG
Sudermannstr.
6700 Ludwigshafen

Prof. Dr. G. Segre
Universita di Siena
Instituto di Farmacologica
Via Banchi di Sotto 55
I 5300 Siena / Italien

Dr. Johann Schäffer
EMD der GSF München
Arabella 4/I
8000 München

Dr. Dieter Schenzle
Inst. f. Med. Biometrie
Äulestr. 2
7400 Tübingen

Prof. Dr. Berthold Schneider
Med. Hochschule Hannover
3000 Hannover

Dr. K.T. Schürger
DKFZ
Inst. f. Dok., Inform. u. Stat.
Im Neuenheimer Feld 280
6900 Heidelberg 1

Alexander Schütt
Inst. f. Med. Dok. u. Stat. der
Universität Köln
Joseph-Stelzmann-Str. 9
5000 Köln 41

Dr. Volker Steinijans
BYK Gulden
GYK-Gulden-Str. 2
7750 Konstanz

Prof. Dr. P. Tautu
DKFZ
Inst. f. Dok., Inform. u. Stat.
Im Neuenheimer Feld 280
6900 Heidelberg 1

Dr. Karl-Heinz Tews
Klinikum der Univ. Frankfurt
Abt. f. Biomathematik
Theodor-Stein-Kai 7
6000 Frankfurt

Dr. Hans-Joachim Trampisch
AST. Biomath. in FB 18
der Universität Gießen
Frankfurter Str. 100
6300 Gießen

Prof. Dr. N. Victor
Univ. Gießen
Frankfurter Str. 100
6300 Gießen

Dr. Wahrendorf
DKFZ
Inst. f. Dok., Inform. u. Stat.
Im Neuenheimer Feld 280
6900 Heidelberg 1

Dr. Walleitner
IABC
Einsteinstr.
8012 Ottobrunn

Prof. Dr. E. Weber
DKFZ
Inst. f. Dok., Inform. u. Stat.
Im Neuenheimer Feld 280
6900 Heidelberg 1

Dr. Gerhard Weckesser
Inst. f. Med. Dok., Stat. u.
Datenverarbeitung
Im Neuenheimer Feld 325
6900 Heidelberg

H. Erich Wichmann
Med. Univ.-Klinik Köln
Abt. Dipl.-Phys.
Joseph-Stelzmann-Str. 9
5000 Köln 41

Dipl.-Math. R. Zentgraf
DKFZ
Inst. f. Dok., Inform. u. Stat.
Im Neuenheimer Feld 280
6900 Heidelberg 1

Table of Contents

III. Pharmacokinetics Chairman: R. Repges page

THE INCIDENCE OF INFECTIOUS DISEASES UNDER THE INFLUENCE OF SEASONAL FLUCTUATIONS

K. Dietz
Health Statistical Methodology
World Health Organization, 1211 Geneva 27, Switzerland

INTRODUCTION

The present paper is concerned with the effect of seasonal variations of the contact rate on the incidence of infectious diseases. The regular oscillations of the number of cases around the average endemic level has attracted the attention of epidemiologists and mathematicians alike. In particular, the two-year period of measles in some large communities has been the object of many attempts of explanation in terms of deterministic and stochastic models. Soper's [1] deterministic approach produced damped oscillations in contrast to the observations. Bartlett [2] suggested that a stochastic version of Soper's model was more realistic. (See also Bailey [3], Chap. 7.) London and Yorke [4] however were able to obtain undamped oscillations with periods of one and two years using a deterministic model which includes a latent period between the time of infection and the beginning of the infectious period. From their simulations they conclude that the length of the latent period has to be within a small range for the occurrence of biennial outbreaks. Recently, Stirzaker [5] treated this problem from the point of view of the theory of nonlinear oscillations according to which the biennial cycles of measles epidemics could be understood as subharmonic parametric resonance.

All the models mentioned above agree in the following assumptions ; (a) susceptibles are added to the population at a constant rate, (b) the death rates of susceptibles and infectives are zero. This latter assumption leads to an equilibrium value for the number of infectives which is independent of the contact rate. If one introduces however a positive death rate, then the contact rate has to exceed a critical level in order for a positive endemic level to exist. The critical level corresponds to the threshold condition for an epidemic to occur in a closed population where no new susceptibles are added. In this paper the behaviour of this modified model is studied under the influence of seasonal variations of the contact rate. Some of the results were already included in a paper read at the Conference on Epidemiology, Alta, Utah, July 8-12, 1974 [6], but they are repeated here for ease of reference.

THE MODEL WITH CONSTANT CONTACT RATE

Let x, y, z denote the number of susceptibles, infectives and immunes, respectively. We assume that individuals have a constant age-independent death rate μ which is also independent of their epidemiological state. The latter assumption implies that we disregard deaths caused by the infectious agent under consideration. The birth rate is assumed to be equal to the death rate so that the population size $n = x + y + z$ remains constant. It is assumed that the number of contacts per unit of time per individual with other individuals in the population is proportional to the size of the population. The constant of proportionality is denoted by β. A contact is understood to be sufficient to cause an infection if a susceptible meets an infective. The constant β is therefore dependent on the type of infection. After an effective contact with an infective, a susceptible becomes immediately infective himself and recovers at a rate γ. After recovery, an individual stays lifelong immune. These assumptions lead to the following set of differential equations

$$\frac{dx}{dt} = n\mu - (\beta y + \mu)x,$$

$$\frac{dy}{dt} = \beta yx - (\gamma+\mu)y, \qquad (1)$$

$$\frac{dz}{dt} = \gamma y - \mu z.$$

The term βyx is derived from the product $\beta n\,(y/n)x$, i.e. the rate at which susceptibles are infected equals the product of the number of contacts they have per unit of time (βn) times the proportion of those contacts which are infective (y/n). This form of the infection rate corresponds to the law of mass action in the theory of chemical reactions. Let u,v,w be the proportions of susceptibles, infectives and immunes, respectively, i.e.

$$u = x/n; \quad v = y/n; \quad w = z/n. \qquad (2)$$

Then the system (1) has the following equilibrium solutions : $(u_0, v_0, w_0) = (1, 0, 0)$ and, for $R = \beta n/(\gamma + \mu) > 1$,

$$u_1 = 1/R$$

$$v_1 = (1 - 1/R)/M \qquad (3)$$

$$w_1 = (1 - 1/R)(1 - 1/M),$$

where $M = (\mu + \gamma)/\mu$.

The quantity R can be called the reproduction rate of the infection since it represents the number of secondary cases that one infective case can produce if introduced into a population of n susceptibles throughout his infectious period. In order for (u_1, v_1, w_1) to be stable, the reproduction rate has to exceed one, which is intuitively obvious. It is also easy to understand that the equilibrium proportion of susceptibles equals $1/R$, since at equilibrium each case should produce on the average one secondary case, i.e. $Ru = 1$, where Ru is the actual reproduction rate in a population where only the proportion u is susceptible. (The proportion $1-u$ of contacts is "wasted" on nonsusceptibles.)

In order to reduce the present model to those mentioned above, one has to let μ tend to zero, n tend to infinity, so that μn tends to some constant σ, where σ is the rate at which susceptibles are added to the population. In the limit, the reproduction rate R tends to infinity, $u_1 = v_1 = 0$, $w_1 = 1$, but $x = \gamma/\beta$ and $y = \sigma/\gamma$.

In [6] a simple relationship was derived between R and the average age A at which individuals contract the infection. If $L = 1/\mu$ denotes the average life length of an individual, then

$$R = 1 + L/A. \tag{4}$$

The average age A at infection is the inverse of the "force of infection" λ which is estimated by fitting the so-called "simple catalytic curve" to the age-specific prevalence of the immunes (see Muench [7]). For example Berger [8] estimates λ (or a in his notation) for rubella to be 0.095 per year. This corresponds to an average age at infection of about 10.5 years. If one assumes an average life length L of 70 years, then (4) yields a reproduction rate for rubella of 7.65. It should be emphasized that this estimation method is based on the assumption that the equilibrium endemic level did not change during the life span of the oldest members included in the sample.

The effect of a vaccination programme can easily be incorporated into the present model. Let s be the proportion of susceptibles who are vaccinated when they are added to the population. (For simplicity we could assume that vaccination takes place at birth.) Then (1) is replaced by the following set of equations :

$$\begin{aligned} \frac{dx}{dt} &= n\mu(1-s)-(\beta y + \mu)x, \\ \frac{dy}{dt} &= \beta yx -(\gamma + \mu)y, \\ \frac{dz}{dt} &= \gamma y - \mu z, \\ \frac{dz'}{dt} &= n\mu s - \mu z', \end{aligned} \tag{5}$$

where z' is the number of vaccinated individuals. It is assumed that the vaccine gives lifelong protection. If we introduce the variables $u' = x/n'$; $v' = y/n'$, $w' = z/n'$, where $n' = n(1-s)$ for $s < 1$, then we get the same equilibrium solutions as (3) with R replaced by $R' = R(1-s)$. The condition $R' < 1$ yields a minimum requirement for the vaccination coverage :

$$s > 1 - 1/R. \tag{6}$$

Thus, e.g. if one were to eradicate rubella with R = 7.65, then one would have to vaccinate at least about 87%. The condition (6) has been stated by Smith [9] without spelling out the underlying model. See [6] for a discussion on the estimation of R from the proportion of susceptibles at the end of an epidemic.

SEASONAL VARIATIONS OF THE CONTACT RATE

We now assume that the contact rate β undergoes a simple harmonic oscillation with a period of one year. Thus we replace the constant β in (1) by the function

$$\psi(t) = \beta(1 - b\cos\omega t). \tag{7}$$

It is assumed that $b < 1$ since otherwise the contact rate would be negative. We have chosen the minus sign in (7) so that at the beginning of the year the contact rate is always at its minimum. If the prevalence of the infection is in phase with the contact rate, then this corresponds to the recommendation by Anderson [10] to define the epidemiologic year as the period between the minima of the prevalence. We shall see later that a considerable phase shift may occur between the peak contact rate and the peak prevalence. If we were to follow Anderson's recommendation in those cases we would have to shift the beginning of the year when chaning β and b.

We introduce the deviations from the equilibrium proportions :

$$\eta = v_1 - v; \quad \xi = w_1 - w. \tag{8}$$

After some straightforward calculations, which are not reproduced here, one obtains the following second order differential equation for ξ

$$\begin{aligned}\ddot{\xi} + &\{\mu R - [\mu(R-1)-(\gamma+\mu)]b\cos\omega t\}\dot{\xi} + \\ &\{\mu(\gamma+\mu)(R-1) - \mu(\gamma+\mu)(R-2)b\cos\omega t\}\xi = \\ &\quad -(1-1/R)\mu\gamma b\cos\omega t - \\ R(1+\mu/\gamma)&(1 - b\cos\omega t)[\mu(\mu+\gamma)\xi^2 + (\gamma+2\mu)\dot{\xi}\xi + \dot{\xi}^2],\end{aligned} \tag{9}$$

where the dots denote derivatives with respect to t. If one neglects the nonlinear terms on the right hand side, one is left with a non-homogeneous linear equation with periodic coefficients. In principle one could apply the asymptotic methods of the theory of nonlinear oscillations for small amplitudes b (see, e.g. [11]). These lead however quickly to very tedious calculations and would only be valid in a limited range of b. In order to gain some information about the stable oscillations of (1) under the influence of the contact rate (7) a number of numerical solutions were obtained.

At first we determine the eigenfrequency of the system in the absence of a seasonal oscillation. For b = 0, the linear part of Eq. (9) reduces to

$$\ddot{\xi} + \mu R\dot{\xi} + \mu(\gamma+\mu)(R-1)\xi = 0. \tag{10}$$

The solution of (10) has the frequency ν with

$$\nu^2 = \mu(\gamma+\mu)(R-1) - \mu^2R^2/4 \tag{11}$$

provided that

$$2M(1-\sqrt{1-1/M}) < R < 2M(1+\sqrt{1-1/M}). \tag{12}$$

Since for the diseases under consideration the human life expectancy is much longer than the infectious period, inequality (12) can be approximated by

$$1 + 1/4M < R < 4M-1. \tag{13}$$

If the frequency ω of the contact rate is close to ν then we expect simple resonance, if $\nu \approx \omega/2$, then we expect subharmonic resonance. Since μ is small we can calculate approximately the reproduction rate $\underline{R}$ for which the resonance oscillation has period $\underline{N}$ years, $\underline{N} = 1,2$, for given γ and μ:

$$R \approx 1 + \omega^2/\left[N^2\mu(\gamma+\mu)\right]. \tag{14}$$

If we eliminate $\underline{R}$ from (4) and (14) we obtain

$$A \approx N^2(\gamma+\mu)/\omega^2. \tag{15}$$

Let us consider the special case of $\underline{N} = 2$, i.e. biennial outbreaks. Since $\underline{I} = 1/(\gamma+\mu)$ is the length of the infectious period, we get

$$A \approx 1/\pi^2 I, \tag{16}$$

if $\underline{I}$ is expressed in years. Here are some examples for illustration

I							
I [years]	0.01096	0.01370	0.01644	0.01918	0.02192	0.02466	0.02740
I [days]	4	5	6	7	8	9	10
A [years]	9.25	7.37	6.16	5.28	4.62	4.11	3.70

If one assumes an infectious period for measles of one week (see [12], page 146), then for biennial outbreaks to occur the average age of infection should be around 5.3 years. Using a modified catalytic model Griffiths [13] estimated that the mean age at attack of measles varied in England and Wales from about 6 years in 1956 to about 4.5 years in 1968 (see [13], Fig.3.) Before discussing these estimates later we shall now summarize the results of our numerical studies on system (1) with contact rate (7).

The differential equations were solved by a Runge-Kutta method which adjusted continuously the stepsize in order to keep the error below some specified bound. For all parameter values chosen, the integration was continued until a certain periodic pattern emerged. Depending on the parameters and the initial values chosen it took on the average about 100 yearly cycles to reach a periodic pattern. The approach to a periodic pattern was accelerated by starting with the equilibrium solution (3) corresponding to $\underline{b} = 0$, with a subsequent linear increase of $\underline{b}$ from zero to its maximum level during an initial period of about 20 yearly cycles. In the following we shall only describe the periodic pattern which was eventually established for a given choice of the parameters. We shall only study the prevalence of the infectives as a function of time. The graphs of Figs. 1-5 are based on computer plots which used 24 equidistant time points per year. The rates were therefore expressed in relation to the time unit of 1/24 years corresponding to about 15.2 days. In all runs the human death rate μ was set equal to 0.000625 which is equivalent to a life expectancy of 66.67 years. The frequency ω of the contact rate has always the value $2\pi/24 = 0.2618$.

Figs. 1-4 describe the effect of varying either $\underline{b}$ or $\underline{R}$ for the fixed value $\gamma = 1.7128$ corresponding to an infectious period of 8.9 days. For those values μ, ω and γ, (14) yields $\underline{R} = 65$ and 17 for the reproduction rates which give resonance oscillations of period one and two years, respectively. The corresponding average ages at first attack are 1.04 and 4.16 years.

In Fig. 1 we keep $\underline{b}$ fixed at 1% and give $\underline{R}$ the values 17, 37, 50, 65 and 82. ($\underline{R}$-1 are the squares of 4,6,7,8 and 9.) As we expect, the amplitude of the prevalence is largest at $\underline{R} = 65$, i.e. when $\omega = \nu$. A 1% amplitude in the contact rate produces a 39.4% amplitude in the prevalence. (The amplitude in the prevalence is calculated as $[v(\max) - v(\min)]/[v(\max) + v(\min)]$. At $\underline{R} = 65$, the contact rate and the prevalence are in phase. For $\underline{R} < 65$, the peak in the prevalence lags behind the peak in the contact rate, for $\underline{R} > 65$ we observe the opposite. If we compare the timings of the peaks in the two cases $\underline{R} = 17$ and $\underline{R} = 82$, we find that they are about 5 months apart. This is quite remarkable considering that the underlying cause is the same in the two cases. From the pattern of seasonal variations, epidemiologists try to deduct the factors which influence the transmission of a particular infectious disease. Differences in the pattern have led sometimes to the conclusion that the underlying mechanisms must also differ (see [10], p. 890, and [14], pp. 277-290). The present example shows that this is not necessary.

In Fig. 2 we keep $\underline{R} = 65$ fixed and increase the amplitude of the contact rate to take the values 2.5%, 5% and 10%. The resulting amplitudes in the prevalence are 74%, 90% and 97%, respectively. The peak prevalence undergoes a phase shift to the right. For $\underline{b} = 10\%$ the prevalence peak occurs two months later than the peak of the contact rate. These graphs show that relatively small variations in the contact rate can cause very large variations in the prevalence, so large that they could not be

FIG. 1 THE EFFECT OF b=0.01 AT DIFFERENT LEVELS OF R (I=8·9 DAYS)

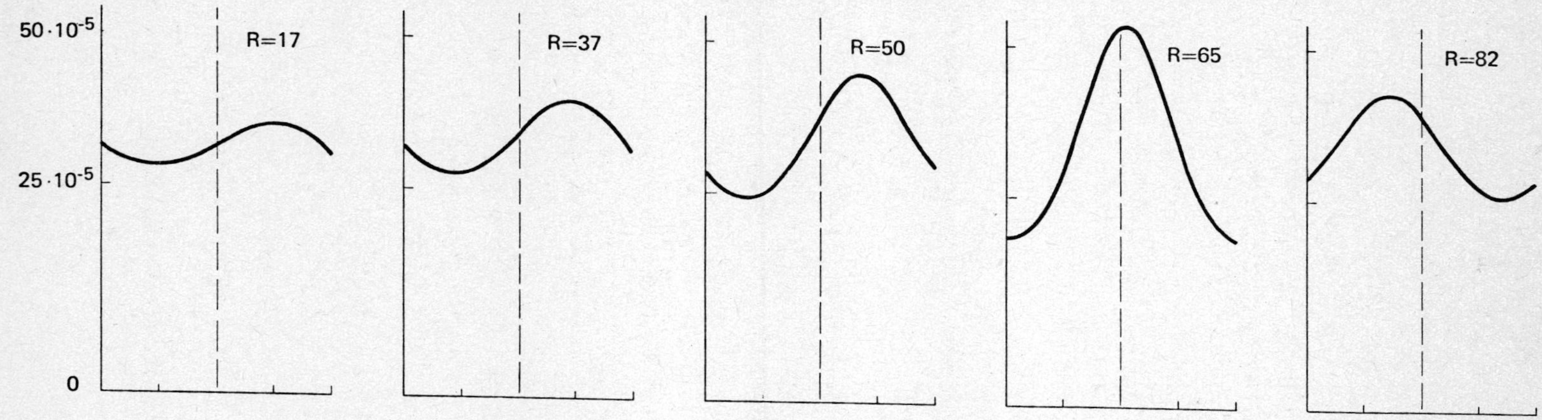

FIG.2 THE EFFECT OF VARYING b AT R=65 (I=8·9 DAYS)

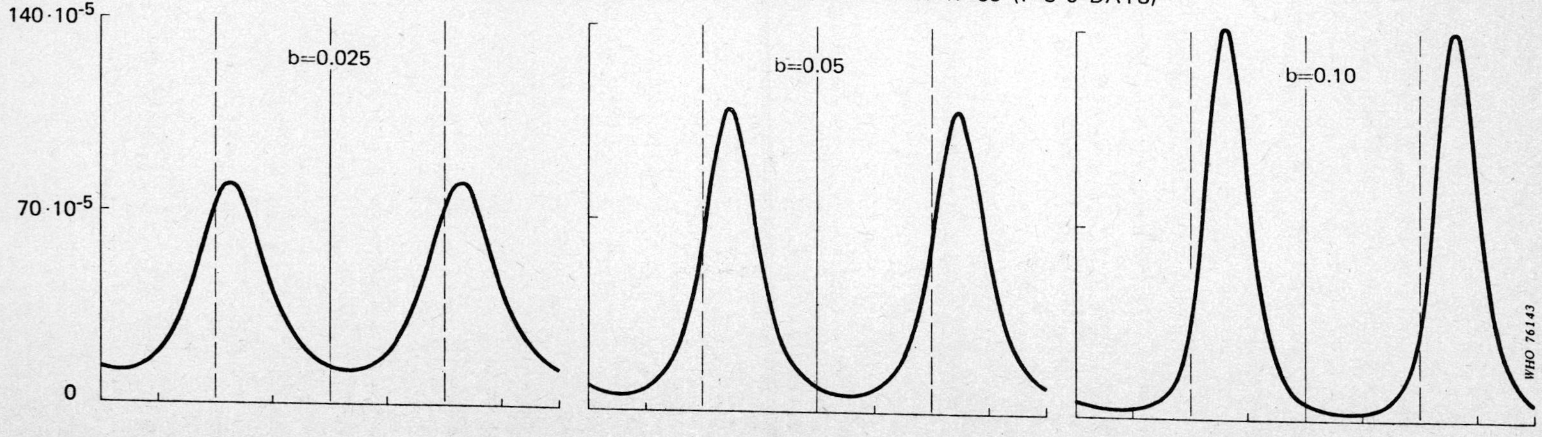

maintained regularly in populations of less than a million, since the minimum values get so small that the chance of extinction would be very high in a more realistic stochastic model.

Now we keep again the amplitude constant, but at 5%. We let R go through a set of values in the vicinity of R = 17, where we expect a biennial resonance oscillation. For R = 14.44 we have a biennial pattern with two nearly equal annual components(Fig.3). For the slightly larger value R = 14.75, the two annual components are markedly different. At R = 15.69, the lower peak is hardly noticeable. As R increases, the lower peak disappears and the higher peak increases until a critical range is reached where the two-year pattern quickly reverts to a one-year pattern. For R = 21.25 the two peaks are again nearly equal and the amplitudes of the prevalence are about 9 times as large as the amplitude in the contact rate. These runs show that the minimum amplitude in the contact rate which is required to generate biennial outbreaks has a value between 0.01 and 0.05. The amplitude of the biennial outbreak is not a symmetric function of the difference between the eigenfrequency of the system and the biennial resonance frequency. For increasing R the amplitude gradually increases but then suddenly drops to nearly the same value at which the two-yearly behaviour started to emerge.

Increasing the amplitude even more to 10% widens the range of R values for which biennial outbreaks occur (Fig.4.) For R = 22.78 we have a one-year pattern with b = 0.05, but at b = 0.10 we get a pattern of two-yearly outbreaks which repeats itself only every six years. Again, the minima between the outbreaks are so low that extinction would be likely in a stochastic model.

According to (14) one could expect patterns with period 3 and 4 in the vicinity of R = 8.1 and R = 5 respectively. For amplitudes up to 10% only one-year patterns were found. Larger amplitudes in the contact rate produced irregular patterns which were not followed up further.

We shall now describe the effect of changing γ which determines the length of the infectious period I. According to (14), a descrease in γ, i.e. an increase in I, increases the reproduction rate at which resonance occurs. E.g. setting $\gamma = 1$ (i.e. I = 15.2 days) gives an R value of 28.4 for a biennial resonance to occur. This increase in R is also coupled with an increase in the minimum value of b for which a biennial pattern emerges. This is to be expected in view of the fact that according to (9) the amplitude of the periodic coefficient of ξ is proportional to $b(\gamma+\mu)$. Fig. 5 shows the resulting behaviour for R = 40 with three different levels of b. At 10% amplitude of the contact rate we have still a yearly pattern of outbreaks. For b = 20% there is a strong two-year pattern and for b = 30% we even get a four-year pattern.

FIG. 3 THE EFFECT OF b=0.05 AT DIFFERENT LEVELS OF R (I=8·9 DAYS)

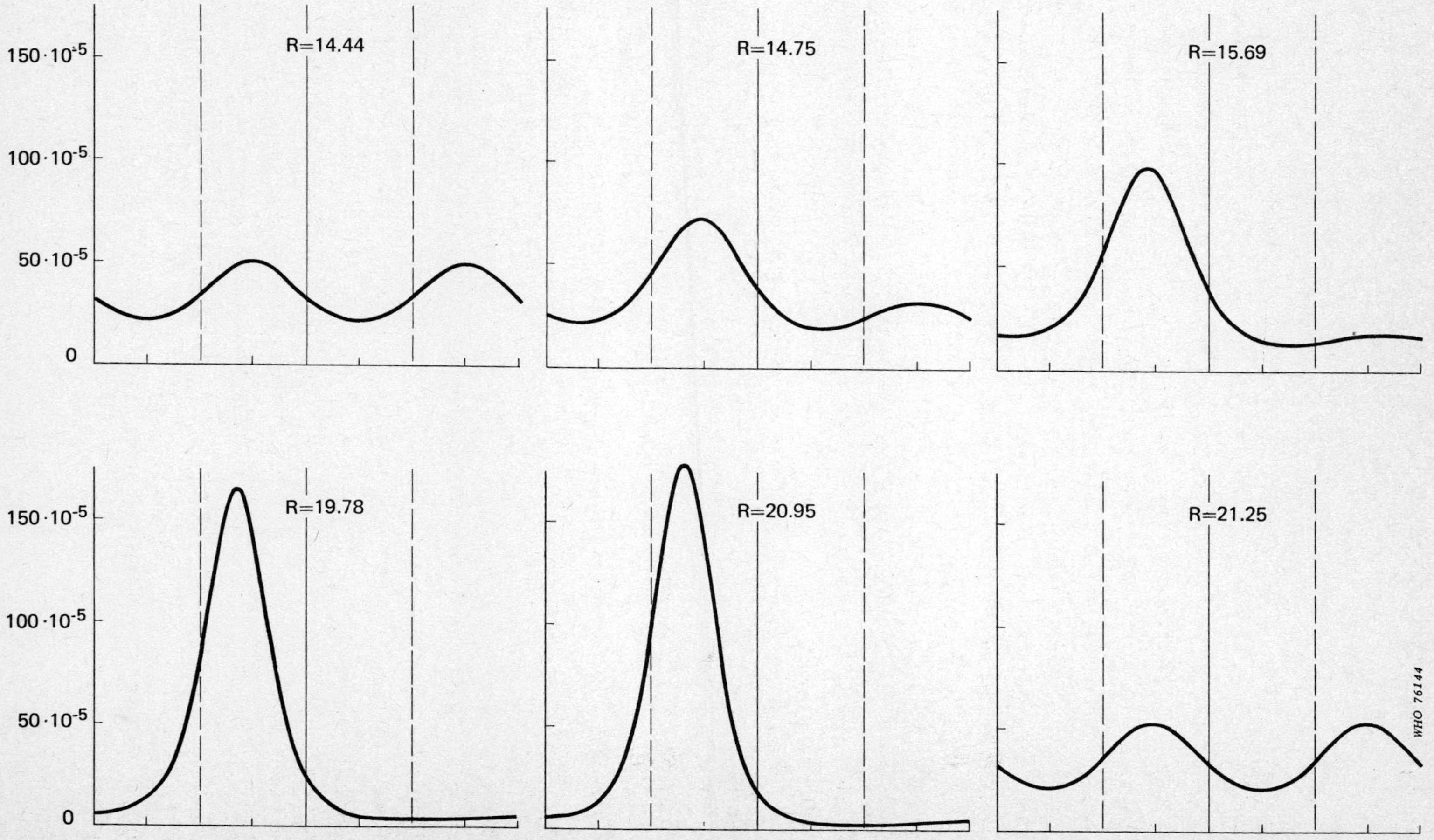

FIG. 4 THE EFFECT OF b=0.10 AT TWO LEVELS OF R (I=8·9 DAYS)

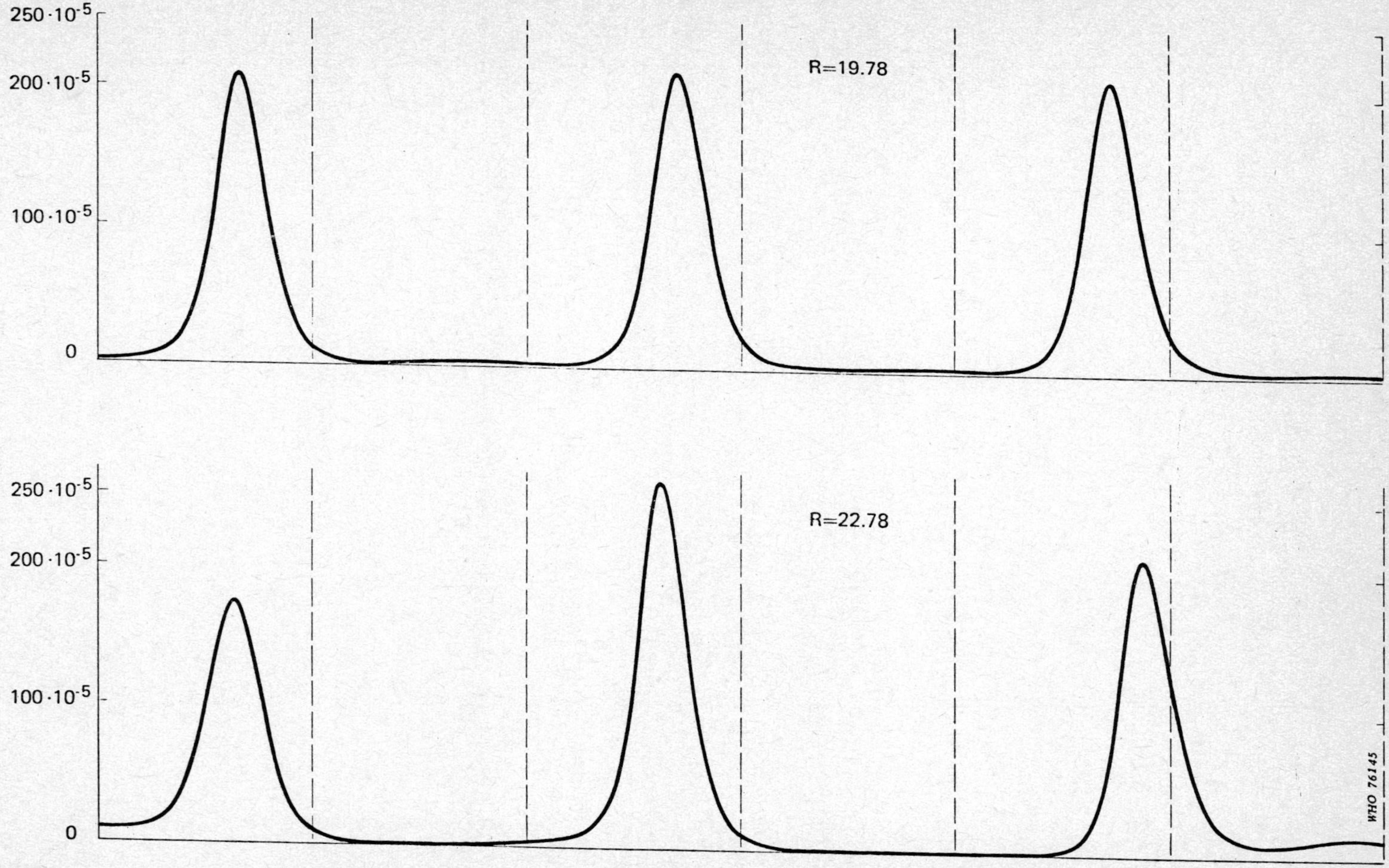

FIG. 5 THE EFFECT OF VARYING b AT R=40 (I=15.2 DAYS)

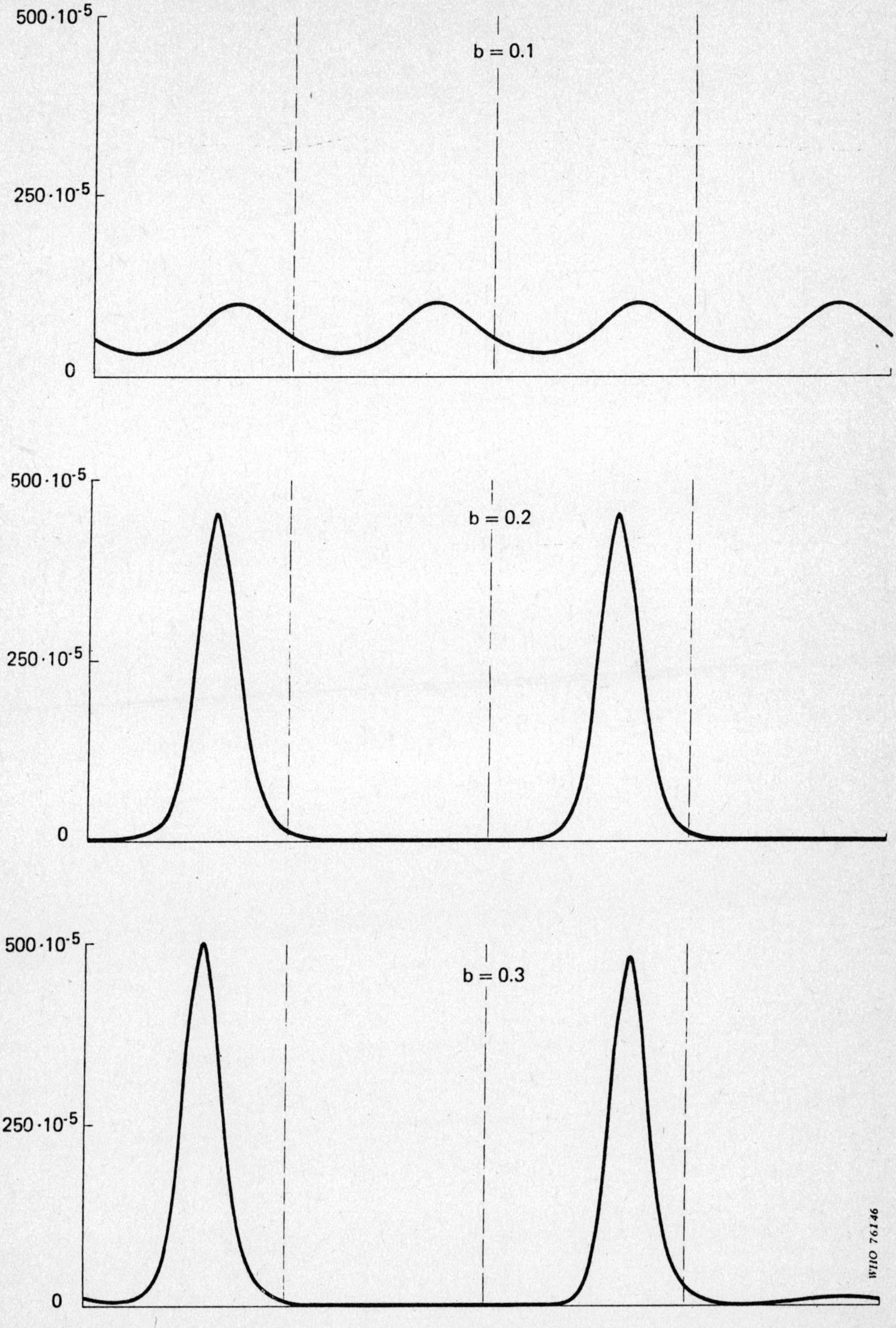

Some runs were also made with $\gamma = 4$ ($\underline{I} = 3.8$ days) which are not reproduced since they showed no new behaviour. The resonance values for $\underline{R}$ and the minimum amplitudes required for biennial outbreaks are smaller than the corresponding values for $\gamma = 1.7128$.

EFFECT OF LATENT PERIOD

Many infectious diseases can only be transmitted from one host to another after a certain latent period. Some models take this fact into account and introduce a delay into (1). We shall modify (1) by introducing a latent period with a negative exponential distribution (see [3] p. 130). Thus (1) is replaced by the following system :

$$\begin{aligned} \frac{dx}{dt} &= n\mu - (\beta y_2 + \mu)x, \\ \frac{dy_1}{dt} &= \beta y_2 x - (\delta+\mu)y_1, \\ \frac{dy_2}{dt} &= \delta y_1 - (\gamma+\mu)y_2, \\ \frac{dz}{dt} &= \gamma y_2 - \mu z, \end{aligned} \tag{17}$$

where $\underline{y_1}$ is the number of individuals who are in the latent period, i.e. they have been infected but are not yet infective. The variables $\underline{x}$, $\underline{y_2}$ and $\underline{z}$ in (17) have the same definitions as $\underline{x}, \underline{y}, \underline{z}$ in (1). The rate δ determines the transition rate at which individuals pass from the latent state into the infectious state. The model (1) corresponds to an infinite δ. The formula for the reproduction rate is now

$$R = [\beta n/(\gamma+\mu)]\,[\delta/(\delta+\mu)]. \tag{18}$$

The nontrivial equilibrium solution of (17) is given by

$$\begin{aligned} x/n &= 1/R, \\ y_1/n &= (1-1/R)(\gamma+\mu)/(\delta C), \\ y_2/n &= (1-1/R)/C, \\ z/n &= (1-1/R)\gamma/(\mu C), \end{aligned} \tag{19}$$

where $\underline{C} = 1 + \gamma/\mu + (\gamma+\mu)/\delta$.

We shall give an approximate formula for the eigenfrequency ν of the system (17) if perturbed around the equilibrium (19). After linearizing (17) and taking into account that the first equation is redundant if we keep the size of the total population at its equilibrium value, we obtain the following cubic equation the solutions of which determine the behaviour of (17) near (19) :

$$\lambda^3 + \lambda^2 (\rho + \delta + \gamma + 3\mu) + \lambda(\rho + \mu)(\delta + \gamma + 2\mu) + \rho(\gamma + \mu)(\delta + \mu) = 0, \tag{20}$$

where

$$\rho = (\gamma + \mu)(1 + \mu/\delta)(R-1)/C. \tag{21}$$

The intermediate steps have not been reproduced here since they are straightforward. In the range of parameter values that is of interest to us, (20) has one real and two complex roots. For the eigenfrequency of (17) we need to know the imaginary part of these complex roots. We rewrite (20) in the following form :

$$\lambda^2 = - \frac{\rho(\gamma + \mu)(\delta + \mu) + \lambda(\rho + \mu)(\delta + \gamma + 2\mu)}{\rho + \delta + \gamma + 3\mu + \lambda} = \tag{22}$$

setting $\lambda = i\nu$ we get

$$\nu^2 = \frac{\rho (\gamma + \mu)(\delta + \mu) + i\nu(\rho + \mu)(\delta + \gamma + 2\mu)}{\rho + \delta + \gamma + 3\mu + i\nu}. \tag{23}$$

If we take into account that μ and ρ are small compared to δ and γ, we can approximate ν^2 as follows

$$\nu^2 \approx \mu(R-1)/(I + D), \tag{24}$$

where $\underline{D} = 1/(\delta + \mu)$ is the average length of the latent period. Hence, the reproduction rate and therefore the age at first attack for which biennial outbreaks occur is a simple function of the sum of the latent and infectious period. In other words, (16) should be replaced by

$$A \approx 1/\left[\pi^2(I + D)\right]. \tag{25}$$

What does this imply in the case of measles ? If we accept that the average age at attack is around 4.5 years, then the occurrence of biennial outbreaks requires that the sum of the latent and infectious period should not exceed about 9 days. This seems to contradict the estimates given in [12]. If we accept a latent period of about a week, the infectious period should only be one or two days in the average. In [12] only a range is given. One must also take into account that a person may be potentially infective, but due to isolation no infection can take place. Before the present model can be fitted to actual data, it should be investigated to what extent the simplifying assumptions made may influence the parameter estimates. For instance, we might ask whether the shape of the distribution of the latent or infectious period affects the resonance behaviour. In this connection, numerical studies showed that the introduction of latent period with a χ_4^2 - distribution gave the same behaviour as model (17). Another factor which might play a role is the age-dependence of contact rates and death rates.

It is hoped that the numerical results presented in this paper stimulate further analytical work on epidemiological models with periodic contact rates. Approximate formulae for the amplitude of the prevalence as a function of R and b would allow to estimate the underlying parameters. This improved understanding of the transmission dynamics could improve the planning of vaccination programmes. It would be interesting to make the proportion s of susceptibles vaccinated in (5) also a periodic function of time with the same frequency ω as the contact rate, but with a different amplitude and phase.

ACKNOWLEDGEMENT

The author would like to thank Mr A. Thomas for the programming required for the drawing of the Figures.

REFERENCES

1. SOPER, H.E. "Interpretation of periodicity in disease prevalence", J.R.Statist.Soc., 92 (1929), 34-73.

2. BARTLETT, M.S. "Deterministic and stochastic models for recurrent epidemics", Proc.Third Berkeley Symp.Math.Statist.& Prob., 4 (1956), 81-109.

3. BAILEY, N.T.J. The Mathematical Theory of Infectious Diseases and its Applications, (2nd edn) Griffin, London and High Wycombe (1975).

4. LONDON, W.P. and YORKE, J.A. "Recurrent outbreaks of measles, chickenpox and mumps. I : Seasonal variation in contact rates", Amer.J.Epidem., 98 (1973) 453-468.

5. STIRZAKER, D.R. "A perturbation method for the stochastic recurrent epidemic", J.Inst.Maths Applics, 15 (1975), 135-160.

6. DIETZ, K. "Transmission and control of arbovirus diseases", in Proceedings of a SIMS Conference on Epidemiology, Ludwig, D. and Cooke, K.L., eds., SIAM, Philadelphia, (1975), 104-121.

7. MUENCH, H. Catalytic Models in Epidemiology, Harvard Univ. Press (1959).

8. BERGER, J. "Zur Infektionskinetik bei Toxoplasmose, Röteln, Mumps and Zytomegalie, Zbl.Bakt.Hyg., I Abt.Orig. A, 224 (1973), 503-522.

9. SMITH, C.E.G. "Prospects for the control of infectious disease", Proc.roy. Soc. Med., 63 (1970), 1181-1190.

10. ANDERSON, G.W. "The principles of epidemiology as applied to infectious diseases" in Bacterial and Mycotic Infections of Man (4th edn), Dubos, R.J. and Hirsch, J.G. eds., Lippincott, Philadelphia (1965), 886-912.

11. SCHMIDT, G. Parametererregte Schwingungen, VEB Deutscher Verlag der Wissenschaften, Berlin (1975).

12. BENENSON, A.S. (ed.), Control of Communicable Diseases in Man (11th edn) American Public Health Association, Washington, D.C. (1970).

13. GRIFFITHS, D.A. "A catalytic model of infection for measles", Appl.Statist., 23, (1974), 330-339.

14. TAYLOR, I. and KNOWELDEN, J. Principles of Epidemiology, (2nd edn). Churchill, London (1964).

SOME SIMPLE MODELS OF THE POPULATION DYNAMICS OF EUCARYOTIC PARASITES

BY

ROY M. ANDERSON

KING'S COLLEGE, LONDON UNIVERSITY,
STRAND, LONDON WC2R 2LS

INTRODUCTION

In recent years our understanding of the processes which control and influence the dynamics of animal populations and lead to their often observed stability and resilience, has in certain areas been enhanced by theoretical studies of a mathematical nature (May, 1973).

The field of epidemiology in particular, is one in which deterministic and stochastic models have greatly aided in the study of the growth and decay of populations of disease organisms. (McKendrick, 1926; Bailey, 1957; Dietz, 1967). Epidemic models in general, describe temporal changes in the number of susceptible, infected and recovered or immune hosts within a population, and depend on a few parameters which specify the nature of the incubation and infectivity periods and the rate of transmission of the disease. The majority of such models are concerned with the dynamics of the spread of procaryotic disease organisms, but research attention has begun to be focused on the factors controlling the population dynamics of eucaryotic parasites; in particular protozoon and helminth diseases.

Theoretical studies of eucaryotic host-parasite systems have been primarily concerned with the dynamics of malaria (Macdonald, 1957), schistosomiasis (Macdonald, 1965; Goffman and Warren, 1970, Cohen, 1970; Nasell and Hirsch, 1972, 1973) and various helminth parasites of domestic animals (Tallis and Leyton, 1966, 1969; Leyton, 1968). More general mathematical and numerical studies of the dynamics of parasite populations have been reported by Kostitzin (1934) and Anderson (1974).

When assessing the biological importance or relevance of many of these studies, difficulties arise at present, due to the paucity of detailed long term investigations into the behaviour of host-parasite population interactions. Furthermore, very few quantitative studies have been made of the many rate parameters which control the flow of protozoon and helminth parasites through their often highly complex life cycles which sometimes involve more than one host species.

The aim of this present paper is to consider in general terms the biology and structure of eucaryotic life cycles, (in particular direct parasite life cycles involving a single species of host) in terms of the dynamics of the various host and parasite populations involved. In essence this is an approach which differs from the more usual epidemiological one, since attention is focused not on the number of susceptible or infected hosts, but on the dynamics of each population involved in the host-parasite interaction. The aim of model formulation is to provide a mathematical framework for investigating the rich dynamical properties which these complex population interactions possess and to outline the processes which aid in the regulation and enhance the stability of such systems.

The models do not aim initially at realism in detail but to provide conceptual insights into the broad classes of behaviour exhibited by these types of population associations. If a model predicts important features which are subsequently found to conform to reality, it will require further study and elaboration. It is regarded as essential, however, for such studies to maintain close links with biological reality by striving to encorporate assumptions which have meaning in the real world and parameters for which estimates can be obtained.

REGULATION OF HOST-PARASITE POPULATION INTERACTIONS

The simplest deterministic model of the interaction between populations of hosts and parasites is that of Lotka and Volterra (c.f. Lotka, 1925). This model assumes that the more abundant the host population (H_t) becomes, the more opportunities there are for the parasite population (P_t) to increase, but as the parasites multiply, the number of hosts destroyed by the parasites increases and thus regulation is achieved. Thus

$$\frac{dH_t}{dt} = (a_1 - b_1P_t)H_t$$

$$\frac{dP_t}{dt} = (- a_2 + b_2H_t)P_t$$

where a_1, a_2, b_1, and b_2 are constants > 0. The behaviour of these equations is well known since they predict oscillations in both host and parasite populations of constant amplitude determined by the initial conditions.

These equations are rather simplistic since in the absence of the parasite, the host population will grow or decay exponentially depending on the parameter values. Leslie and Gower (1960) and Pielou (1969) have suggested alternative models in which the growth of the host population is constrained by density dependent processes in the absence of parasitic infection, and grows according to the logistic pattern first suggested by Pearl and Reed (1920). The constraints are assumed to be created by intraspecific competition for environmental resources. In models of this form both H and P exhibit dampened oscillations in time to an equilibrium point stable to small perturbations.

The majority of protozoon and helminth parasite life cycles are of a more complex structure than the framework suggested by such models. The life cycles of these parasitic organisms often involve more than one species of host and many distinct parasite populations occupying various ecological niches. They can be classified into two general categories; direct cycles involving a single host species and indirect cycles involving two or more species of host. The host within or on which the parasite becomes adult and reproduces sexually is usually termed the definitive or final host, while hosts in which either asexual or no reproduction occurs are termed intermediate hosts.

The number and type of distinct parasite and host populations, and the rate parameters controlling the flow of parasites through a life cycle can be clearly defined by the construction of a diagrammatic population flow chart. This tvpe of qualitative representation can provide the framework for the development of mathematical models and, in addition, helps to clarify the areas in the cycle in which the biological details are poorly understood.

One of the simplest direct life cycles is that exhibited by the common whipworm of man, the nematode Trichuris trichuria (Fig. 1). The parasite develops to sexual maturity in the caecum of its host and when gravid lays eggs which pass to the external habitat with the faeces. The eggs once embryonated act as infective agents and gain entrance to a host by ingestion. Although this cycle is of a very simple form when compared with other protozoon or helminth parasites such as species of Plasmodium or Schistosoma, it involves three distinct populations; the host, adult parasite and eggs. In addition, ignoring the possibility of immigration and emigration in the human population, the flow of parasites through the cycle is controlled by six rate parameters.

The dynamical behaviour of this type of population interaction is very poorly understood, due partly to the complexity of the system and hence the difficulty in formulating appropriate models, and partly to the scarcity of quantitative information concerning the rate parameters involved in such life cycles.

Before examining models describing all the processes outlined in Fig. 1, it is worth briefly considering the dynamics of a single host system. An important concept to examine initially is the effect of parasite induced host mortalities on the stability of the population interaction. For example if it is assumed initially that the parasite does not affect host survival what sort of behaviour do the parasite populations in a direct life cycle interaction exhibit?

The basic features of the adult parasite population P_t within the host, and the egg population W_t in the external habitat are represented by the following equations.

$$\frac{dW_t}{dt} = \lambda_1 P_t - \mu_1 W_t - \lambda_2 W_t$$

$$\frac{dP_t}{dt} = \lambda_2 W_t - \mu_2 P_t \tag{1}$$

where the instantaneous rates are assumed for simplicity to be constant and are defined per unit of time as; λ_1 = rate of egg production per parasite, μ_1 = death rate of eggs per egg, λ_2 = rate of infection per host, μ_2 = death rate of adult parasites per parasite. It is assumed in this model that the rate of infection of the host is directly proportional to the density of the infective stages (eggs).

The solution of these equations demonstrates that both populations either grow or decay exponentially depending on the parameter values and thus no equilibrium state is achieved. Models containing similar assumptions but describing the dynamics of the parasite population within a population of hosts exhibit the same overall pattern of behaviour. Furthermore, even if the host population is regulated by a density dependent constraint of the logistic form, the adult parasite and egg populations will

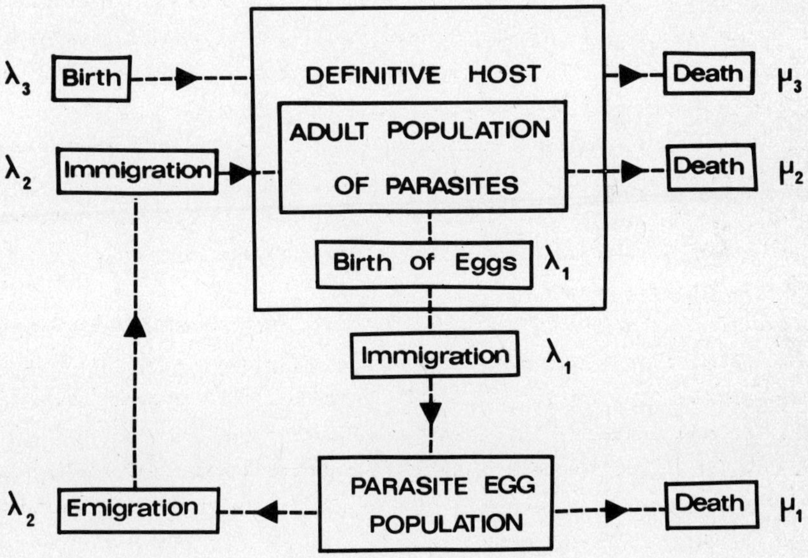

Fig. 1 Diagrammatic representation of a simple direct life cycle (e.g. Trichuris trichuria). The population parameters determining the flow of parasites through the system are defined as rates per individual per unit of time.

λ_1 = egg production rate, μ_1 = egg death rate

λ_2 = host infection rate, μ_2 = parasite death rate

λ_3 = host birth rate, μ_3 = host death rate

not attain a steady state. To achieve an equilibrium, some form of regulatory process either host - induced (immunological) or resulting from intraspecific competition in the parasites micrenvironment, or due to parasite - induced host mortalities must operate on the dynamics of the interaction.

This point can be simply illustrated by again considering the direct life cycle illustrated in Fig. 1. Four population parameters determine the flow of parasites through the cycle and one or more of these rates may be a function of population density. For example, the parameter λ_1, the rate of egg production by adult parasites may vary according to population density, production being inhibited at high densities due to either intra specific competition or as a result of host generated immunological attack. Many reports of this type of response exist in the parasitological literature; for example egg production per fluke by Fasciola hepatica in sheep decreases as the fluke burden increases (Boray, 1969). Similarly Michel (1967) reported that heavy infections of the nematode Ostertagea ostertagi in calves resulted in a decreased egg output per worm when compared with light infections. The death rate of the adult parasite within the host, μ_2, may also be functionally related to density again either due to competition for limited resources or to the hosts immune responses. Jarrett et al (1968) demonstrated that the nature of the immune response to Nippostrongylus muris, a hookworm nematode of rats, depended on the size of the initial dose of larvae administered to the host, large doses resulting in decreased adult worm survival. Similar responses have been reported by Phillips et al (1975) for Schistosoma mansoni infection in rats exposed to various doses of cercariae (Fig. 2).

The precise functional form of the relationship between density of parasites and magnitude of a population rate parameter varies widely in different host parasite interactions and may often be dependent on a temporal dimension such as time from initial exposure.

The broad consequences of such responses on the single host direct life cycle system, however, are easily determined. If it is assumed that the egg production rate λ_1, is a linear function of parasite density such that

$$\lambda_1(P_t) = a - bP_t$$

where the constants a represents the maximum egg production rate and b determines the severity of the density dependent response, then the model exhibits regulated growth to the equilibrium states.

$$\hat{P} = [a - \frac{\mu_2}{\lambda_2}(\mu_1 + \lambda_2)]\ \frac{1}{b}$$

and

$$\hat{W} = \frac{\mu_2}{\lambda_2}\hat{P} \qquad (2)$$

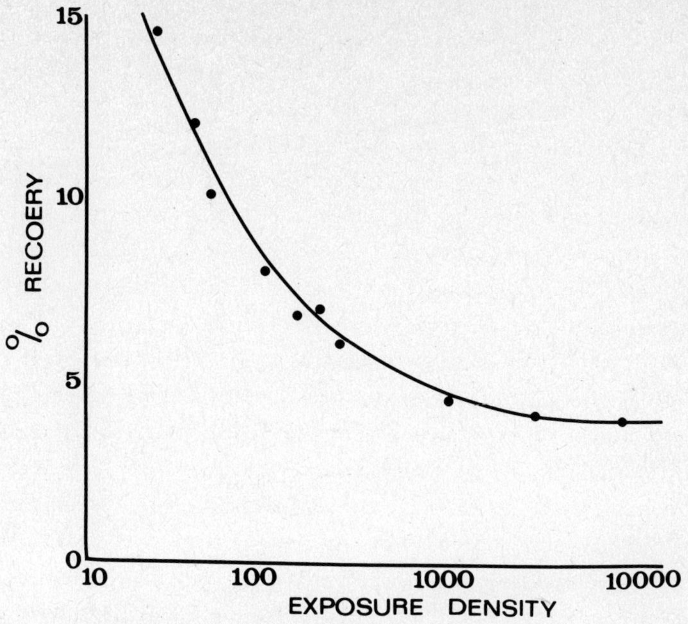

Fig. 2 The relationship between the number of cercariae of Schistosoma mansoni to which a rat is exposed for a constant period of time, and the number of adult worms recovered from the host after a period of six weeks. (% recovery is the % of exposure density recovered). (Data from Phillips et al, 1975).

The positive and thus realistic values of $\hat{P}$ and $\hat{W}$ possess neighbourhood stability. Only certain combinations of parameter values, however, lead to equilibrium states as illustrated in Fig. 3 where the zero isoclines of $\hat{P}$ and $\hat{W}$ are shown in the P_t, W_t phase plane for two different sets of parameter values. Realistic equilibria only exist for intersections of the two isoclines in the positive sector of the plane.

Although in theory many of the rate parameters may be functionally related to parasite density, only a single rate of this form is required to give rise to regulated population growth within the host. It is tempting, however, to draw the speculative conclusion that the more parameters there are of this form in a given life cycle, the more stable the host parasite interaction.

The direct life cycle models outlined so far, are obviously highly oversimplified and in particular they ignore the dynamic aspects of the interaction between host and parasite populations. In many parasite life cycles, the main regulatory influence in the dynamics of the system will be caused by parasite induced mortalities in the host population with the subsequent loss of a proportion of the total parasite population.

The rate of parasite induced host mortality may be an increasing function of parasite population density, or it may be an all or nothing response. This latter type of mechanism was envisaged by Crofton (1971) who considered that the vast majority of parasitic species are capable of killing their hosts if present in large enough numbers. Crofton (1971a and 1971b) also argued that the death of a single host in the tail of an overdispersed distribution of parasite counts per host resulting in the removal of a large number of parasites, was a strong regulatory force in the population interaction. He envisaged the mechanism as operating at a given level and proposed that for a specific host - parasite interaction there existed a lethal burden of parasites, above which the death of the host occurred if further infections were acquired. In reality, the precise density of parasites which result in the death of a host will vary not only with temporal and spatial factors but also within strata of the host population such as age classes. Crofton's concept, however, is still of value since an expected lethal level could possibly be estimated for a given class of hosts.

Some parasite species have sublethal effects, such as the modification of the host's reproductive potential as reported by Pan (1965) for snails of the species _Australorbis glabratus_ infected with larval stages of _Schistosoma mansoni_. All these types of responses, however, will in many life cycles act in conjunction with density dependent constraints on the host population created by intra or inter specific competition.

The direct life cycle can be reformulated to include consideration of the host population in a number of ways. One of the simplest methods of incorporating parasite induced host mortality is to assume that an additional rate operates on the host population, where the death rate γ per host is proportional to the mean parasite burden $\frac{P}{H}$. Thus whatever form of distribution of parasite numbers per host is

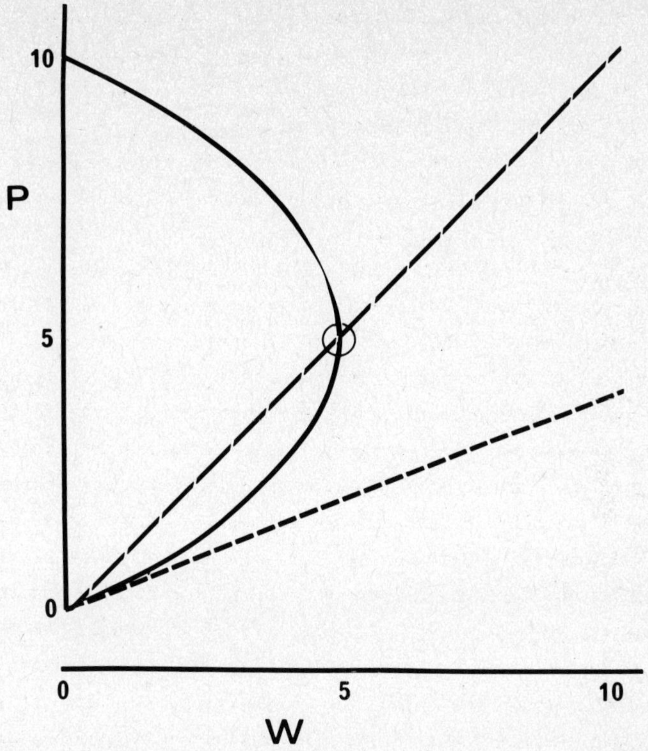

Fig. 3 Phase plane created by the time dependent variables, adult parasite population size P_t and egg population size W_t. Solid line; trajectory predicted by the equation $(a - bP_t)P_t - W_t(\mu_1 - \lambda_2) = 0$ where $a = 1.0$, $b = 0.1$, $\mu_1 = 0.1$, $\lambda_2 = 0.4$ Finely stippled line; trajectory predicted by the equation $(\lambda_2 W_t - \mu_2 P_t) = 0$, where $\lambda_2 = \mu_2 = 0.4$ Stippled line trajectory predicted by the equation $(\lambda_2 W_t - {}_2 P_t) = 0$ where $\mu_2 = 1.0$, $\lambda_2 = 0.4$

generated, an increase in the mean parasite burden will result in a greater proportion of the host population reaching the vicinity of the lethal burden and hence host mortalities will rise. If it is further assumed that, in the absence of parasitic infection, the host population grows according to the logistic model where the host death rate is density dependent and of the form

$$\mu_3(H_t) = \alpha - \beta H_t$$

then a system of equations can be constructed to describe the essential features of the interaction

$$\frac{dW_t}{dt} = \lambda_1 P - \mu_1 W_t - \lambda_2 H W_t \qquad (3)$$

$$\frac{dP_t}{dt} = \lambda_2 . H . W_t - \mu_2 P_t - \mu_3(H_t).P_t - \frac{\gamma P_t^{\,2}}{H_t} \qquad (4)$$

$$\frac{dH_t}{dt} = (\lambda_3 - \mu_3(H_t)\,)\,H_t - \gamma P_t \qquad (5)$$

The host equilibrium population size is given by

$$\hat{H} = \left[- b \pm \sqrt{(b^2 - 4ac)} \right] / 2a \qquad (6)$$

where $a = \lambda_2 \beta$, $b = \lambda_2 (\alpha + \lambda_1 + \frac{\mu_1 \beta}{\lambda_2} - \mu_2 - a)$ and $c = - \mu_1 (\alpha + \mu_2 + a)$;

This expression predicts a single positive and thus realistic steady state for the host population. The remaining equilibria are given by

$$\hat{P} = \left[\lambda_3 - \alpha - \beta \hat{H} \right] \frac{\hat{H}}{a} \qquad (7)$$

$$\hat{W} = \frac{\lambda_1 \hat{P}}{(\mu_1 + \lambda_2 \hat{H})} \qquad (8)$$

The model generates regulated population growth to positive equilibrium states for certain ranges of parameter values. In particular the parameter λ_1, the parasite egg production rate and λ_3 the host birth rate must be sufficiently large to overcome the effects of host and parasite death rates and parasite induced host mortalities.

The stability of this system, with respect to small perturbations can be analysed as described by May (1973).

The community matrix C, or perhaps better termed the life cycle matrix in host parasite models, can be constructed by evaluating all the partial derivatives of the equations 3, 4, and 5 with respect to all three population variables W_t, P_t and H_t at the equilibrium points $\hat{W}$, $\hat{H}$ and $\hat{P}$.

The set of equations describing the time dependence of the perturbations with respect to the equilibrium population values possesses a non trivial solution if and only if

$$\det|(C - x I)| = 0$$

where I is the 3×3 unit matrix and the x's are the eigen values. This condition gives a third order polynomial in x of the form

$$x^3 + z_1 x^2 + z_2 x + z_3 = 0$$

The perturbations will only die away in time if all the eigen values have negative real parts. The Routh - Hurwitz stability criterion, as outlined by May (1973), gives the following conditions for stability

$$z_2 > 0, \quad z_3 > 0 \quad \text{and} \quad z_1 z_2 > 0.$$

In the case of the direct life cycle model these conditions give rise to some rather unattractive expressions, but it is apparent that certain parameter values give rise to equilibrium states which possess neighbourhood stability.

This approach can be used to construct more complex models, describing the population dynamics of indirect life cycles but biological insights become progressively more difficult to obtain from the stability criteria as the number of populations involved and thus number of variables in the model increases.

Deterministic models of this form, however, may be of some use in generating insights into the general effects of implementing various types of epidemiological control programs. The equations obviously oversimplify the complex biological details of parasite life cycles and have inherent weaknesses due to their determinism but they do provide a means of assessing the consequences of altering various rate parameters in the life cycle such as the adult parasite death rate. It may be possible, for example, to increase this particular rate by treatment of the host population with chemotherapeutic agents. An obvious aim in such control programs, would be to move a population parameter into the region of unstable parameter space, where small perturbations of the equilibrium population states lead to the extinction of the parasitic disease.

MULTICOMPARTMENT MODELS

Various types of compartmental models can also be used to examine the dynamics of complex host - parasite interactions. Each individual host within a population for example, can be envisaged as a compartment or box, through which parasites flow at a rate determined by the immigration and death parameters. The hosts are in reality labelled at any one point in time by the number of parasites they contain and thus the total host population can be partitioned into a series of subsets into which individuals are placed according to their parastic burdens.

Thus if $x_i(t)$ is the number of hosts containing i parasites at time t, where $i = 0,1,2,3 \ldots\ldots$, then

$$\sum_{i=0}^{\infty} x_i(t) = H_t \quad .$$

the host population size and P_t the total parasite population is given by

$$P_t = \sum_{i=0}^{\infty} i x_i(t) \quad .$$

The summations can be made over an infinite range if it is assumed that no constraints operate on the build up of parasite numbers within a single host such as the finite size of the parasites' microenvironment. Alternatively, they can be made over a finite range defined by a lethal parasite burden.

This form of notation was first used in connection with host - parasite population models by Kostitzin (1934) whose pioneering work in this area is often ignored by population biologists. Subsequently it has been used by MacDonald (1950) in his treatment of superinfection in humans suffering from malaria and more recently by Bailey (1957) in a similar context. This latter author subdivided the host population into several distinct classes, defined by the number of "broods" of malarial parasites harboured. In Bailey's model the "broods" are regarded as distinct entities within the host. Similar types of notational framework have been used to formulate systems of difference equations to simulate malarial parasite dynamics in human populations divided into various epidemiological classes such as infective, non-infective and immune (Dietz et al, 1974).

The compartmental approach can be used with advantage, to formulate models of the dynamics of specific sections of a parasite life cycle. For example, a model for the immigration - death processes controlling the dynamics of adult parasite populations within a host population of fixed size can be expressed as follows

$$\frac{dx_o}{dt} = -\lambda x_o + \mu x_1$$

$$\vdots$$

$$\frac{dx_i}{dt} = \lambda x_{i-1} + \mu(i+1)x_{i+1} - \lambda x_i - \mu i\, x_i$$

$$\vdots$$

where λ is the constant immigration rate and μ is the death rate per parasite. The solution of these deterministic equations predicts a Poisson distribution of parasites per host at equilibrium, where the proportion of hosts with i parasites is given by

$$P_i = \left[\exp\,(-\lambda/\mu)\;(\lambda/\mu)^i\right] / \; i!$$

This result is to be expected on intuitive grounds, since it is assumed within the model that each host is equally susceptible to infection and thus the aquisition of parasites is essentially the result of chance contacts between host and parasite.

Many parasitic diseases result in an increased or decreased susceptibility in infected hosts with respect to subsequent infections and thus the immigration rate λ is often not constant but dependent on parasite burden. This type of situation arises in the case of the human pin worm Enterobius vermicularis which is parasitic in the intestine of man. The female worms when gravid migrate down the gut of the host to the anal opening and lay eggs on the perianal skin. The host may experience intense irritation in this area and is often stimulated to scratch the infested region. The life cycle of the parasite is direct, the eggs being layed in an embryonated state and thus hand - mouth contacts lead to reinfection.

A more common situation however is for current infections to decrease the probability of future parasites establishing as adult parasites within a host. This type of process may result from intraspecific competition within the parasite's microenvironment, or more generally as a result of immunological attack by the host directed specifically against invading parasitic stages. The rhesus monkey responds in this way to infections of Schistosoma mansoni which is parasitic in the hepatic portal blood system. Adult worms become antigenically disguised and thus survive the host's humoral antibody responses but invading schistosomulae are destroyed en route to the hepatic portal system (Smithers and Terry, 1965).

A simple model can be constructed to examine certain aspects of such responses, in which the infection rate is dependent on the number of parasites already established, by considering a host population of constant size exposed to a finite number of infective larval parasites.

If there are W_0 infective larvae present at the beginning of the process, and the infection rate is of linear form $\lambda_i = \lambda_o + \beta i$ where the constant can be negative or positive depending on the nature of the density dependent infection process, then the following equations represent the basic features of the interaction.

$$\frac{dW_t}{dt} = - W_t \sum_{i=0}^{s} \lambda_i x_i(t) \tag{9}$$

$$\frac{dx_o}{dt} = - \lambda_o x_o(t) W_t \tag{10}$$

$$\vdots$$

$$\frac{dx_i}{dt} = \left[\lambda_{i-1} x_{i-1}(t) - \lambda_i x_i(t)\right] W_t \tag{11}$$

$$\vdots$$

$$\frac{dx_s}{dt} = \lambda_{s-1} x_{s-1}(t) W_t \tag{12}$$

The period of observation of the process is regarded as short and thus births and deaths are assumed to occur in neither host nor parasite populations. The inclusion of the parameter s, provides a finite limit to the system of equations.

When β is positive, susceptibility to further invasion being enhanced by the presence of an established population, the solution of the model gives

$$P_i(t) = \binom{(\lambda_o/\beta) + i - 1}{i} \frac{x_o(t)}{N} \left[1 - (x_o(t)/N)^{\beta/\lambda_o}\right]^i$$

where $P_i(t)$ is the probability of observing i parasites in a single host and is defined as $x_i(t)/N$ where $N = \sum_{i=0}^{s} x_i(t)$.

This expression is very similar to the probability function of the negative binomial distribution (Greenwood and Yule, 1920), and thus the pattern of parasite counts per host is overdispersed (Fig. 4).

Conversely, when β is negative, the more normal situation in the majority of host parasite interactions; the rate of infection being decreased by the presence of established parasites, the solution of the model gives

$$P_i(t) = \binom{\lambda_o/\beta}{i} (1 - Z)^i Z^{(\lambda_o/\beta - i)}$$

where $Z = \left[x_o(t)/N\right]^{\beta/\lambda_o}$.

Negative values of β thus give rise to the probability function of the positive binomial distribution and thus the parasite counts are underdispersed within the host population.

These distributions, generated by density dependent infection rates are of practical interest particularly with respect to laboratory infection experiments. It is important to note, however, that in natural populations of hosts, the many processes which generate overdispersion, such as heterogeneity in exposure to infection, will tend to mask the effects of density dependent processes.

Heterogeneity in exposure to infection is a common feature of many host - parasite interactions, the generating mechanisms being many and varied. The immigration rate λ is thus often a random variable, say Λ. The consequences of encorporating this assumption can most easily be assessed by considering initially a single host containing a parasite population subject to constant immigration and death rates. The stochastic formulation of this process gives rise to the following probability

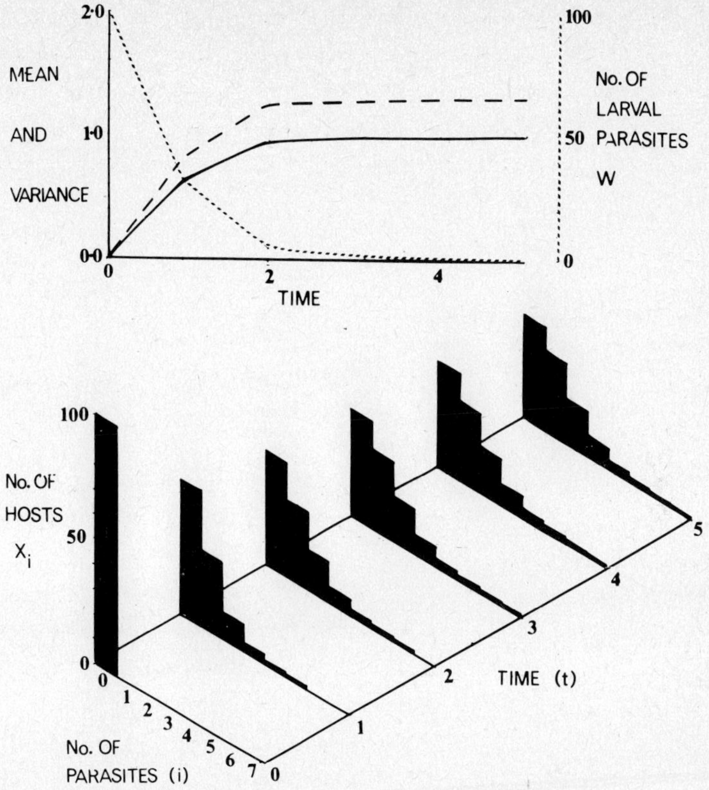

Fig. 4 The dynamics of density dependent infection in a population of hosts. The immigration rate is of linear form where $\lambda_i = \lambda_o + i\beta$. The size of the larval parasite population W_o at t = 0, = 100 and $H_o = \sum_{i=0}^{s} x_i(0) = 100$, $\lambda_o = 0.01$ and $\beta = 0.003$.

<u>Graph A</u> The growth of M, the mean number of parasites per host and the variance σ^2, towards equilibrium. The corresponding decrease in W_t is also illustrated. <u>Graph B</u> The change in time of the negative binomial distribution of the number of parasites per host as the infection process approaches equilibrium.

generating function $\Pi(z,t)$ for $P_n(t)$ the probability of observing n parasites in the host at time t, given the initial conditions $p_o(t) = 1$,

$$\Pi(z,t) = \exp\left[(z - 1)\frac{\lambda}{\mu}[1 - \exp(-\mu t)]\right]$$

the generating function of a Poisson process.

If the immigration rate is now regarded as a random variable Λ, say of Poisson form with mean α and moment generating function

$$M(\theta) = \exp(-\alpha)\exp\left[-\alpha\exp(\theta)\right]$$

then the new generating function $Q(z,t)$ for the probability of observing n parasites at time t is given by

$$Q(z,t) = \exp(-\alpha)\exp\left[-\alpha\exp\left[\frac{1}{\mu}(1 - \exp(-\mu t))(z - 1)\right]\right]$$

This expression is the generating function of the Neyman type A distribution and thus the model predicts the generation of an overdispersed distribution of parasite counts.

Alternatively if the random variable was of a gamma type with moment generating function

$$M(\theta) = (1 - \alpha\theta)^{-\beta}$$

the resulting distribution would be of the negative binomial form. The occurrence of heterogeneity in immigration between members of a host population thus generates a marked degree of overdispersion of parasite population sizes within the host population. As a result of spatial mobility of the individual hosts, this type of pattern of the adult parasite distribution will give rise to spatial heterogeneity in the distribution of infective stages in the free - living habitat.

The examination of segments of a parasite's life cycle is often informative, particularly as stochastic models may be formulated for specific sets of population processes to aid in the estimation of parameter values (Dietz et al, 1974). This approach, however, does not provide a great deal of insight into the dynamical behaviour of the complete host - parasite system. The behaviour of any one segment of a life cycle is dependent to a large extent on the state of the previous segment or population. The dynamics of a complete cycle can thus only be fully understood by regarding all the interconnecting compartments as a single unit possessing its own unique dynamical behaviour.

The compartmental notation can be used to construct such a model, which is in essence very similar to the model described by equations 3,4, and 5. It is assumed in the following model that only single transitions occur between the different classes of the host population in a small interval of time and thus a host containing i parasites can only move to the (i - 1) or (i + 1) classes.

For the majority of species of helminth parasites it is practically feasible to adopt the single transition notation, since the maximum values of s will not prohibit the numerical solution of the model. In the case of protozoon parasites however, where parasite burdens per host may be in the millions alternative definitions must be adopted. For example, the rate parameters can be altered to account for transitions between $x_{i-1} \overset{\leftarrow}{\rightarrow} x_i \overset{\leftarrow}{\rightarrow} x_{i+1}$ classes, where one unit represents hundreds or thousands of parasites rather than a single organism.

The following multicompartment model describes the essential framework of a direct life cycle parasite, where the rate parameters are as defined in Fig. 1. As an initial approach the equations do not encorporate the effects of parasite induced host mortalities, and the hosts are assumed to be able to harbour an infinite number of parasites and thus no finite limit exists for the number of equations describing the population interaction.

$$\frac{dW_t}{dt} = \lambda_1 \sum_{i=0}^{\infty} i\,x_i - \lambda_2 W \sum_{i=0}^{\infty} x_i - \mu_1 W \qquad (13)$$

$$\frac{dx_o}{dt} = \lambda_3 \sum_{i=0}^{\infty} x_i - \lambda_2 W x_o + \mu_2 x_1 - \mu_3 x_o \qquad (14)$$

$$\vdots$$

$$\frac{dx_i}{dt} = \lambda_2 W x_{i-1} - \lambda_2 W x_i + \mu_2 (i+1) x_{i+1} - \mu_2 i x_i - \mu_3 x_i \qquad (15)$$

$$\vdots$$

The non linearities in these equations arise from the infection processes, the $\lambda_2 W x_i$ terms, since it is assumed that the rate or force of infection is proportional to both the number of infective stages in the host's habitat and the number of hosts present. This non linear term is regarded as a basic feature of the biology of host parasite interactions and thus its inclusion in the model is regarded as essential. It does however create some difficulties in achieving analytical solutions for such models.

The behaviour of equations 13, 14 and 15 is easily determined, since both host and parasite populations grow or decay exponentially depending on the values of the birth, immigration and death parameters. If no parasites are present the model collapses to a pure birth and death process for the host population.

The interaction can be regulated in a variety of ways and initially it is worth considering the influence of density dependent responses on both host and parasite populations, in the absence of parasite induced host mortality.

The host population in the absence of parasitic infection is unlikely to grow or decay exponentially and thus a logistic pattern is a more realistic assumption and should be encorporated in the model. A density dependent constraint where the host birth rate is of the form

$$\lambda_3(\sum x_i) = \alpha - \beta \sum x_i$$

can be simply encorporated in the model. Thus the host population will grow either in the presence or absence of the parasitic infection to a stable equilibrium $\hat{H}$ where

$$\hat{H} = (\alpha - \mu_3)/\beta \quad .$$

The inclusion of this constraint on the host population however, does not serve to regulate the parasites and thus a further density dependent mechanism is required as indicated earlier in this paper by the single host model represented in equation 1. This added form of control on the parasite population leads to regulation of both host and parasite populations, but both are to a large extent acting in an entirely independent manner. The parasite population, for example, does not influence the growth of the host population at all, while a single parasite population is simply constrained by the carrying capacity of an individual host and thus the total parasite population is directly proportional to the total number of hosts.

The model obviously requires an interactive aspect which can be encorporated by the inclusion of parasite induced host mortality. In contrast to the direct life cycle models described previously in which no account was taken of the distribution of parasites within the host population, the pathogenic effects of the parasite on the host can be included in the compartmental model in a variety of ways. The host death rate can be formulated as some function of the number of parasites harboured. The simplest functional form would be a linear relationship where

$$\lambda_3(i) = a + bi$$

where a and b are constants.

Alternatively a more interesting and perhaps more realistic way of including this effect in the model is to assume that the vector x_i has a finite limit of magnitude s, where i = 0,1,2.......s. It can then be assumed that the parasites flow through the compartments (hosts) until a value s is reached when both host and the s parasites are removed from the system by the death of the host. This concept as mentioned previously was first proposed in quantitative form by Crofton (1971) and although a rather crude imitation of biological reality it has many attractive features particularly if imagined as an expected value for a given host population.

Interestingly, the inclusion of this assumption in the model, in the absence of any other types of density dependent constraints on either host or parasite population, for a small range of parameter values will give rise to regulated population growth to equilibrium states for all three populations (Fig. 5). The behaviour of the model, however is very sensitive to the initial conditions and rate parameter values.

The inclusion of a density dependent birth rate for the host population such that in the absence of parasitic infection the host population grows to an equilibrium state $\hat{H}$ leads to a better behaved model. The equations become,

$$\frac{dW_t}{dt} = \lambda_1 P - \lambda_2 WH - \mu_1 W \tag{16}$$

$$\frac{dx_o}{dt} = (\alpha - \beta H)H - \lambda_2 Wx_o + \mu_2 x_i - \mu_3 x_o \tag{17}$$

$$\vdots$$

$$\frac{dx_i}{dt} = \lambda_2 Wx_{i-1} - \lambda_2 Wx_i + \mu_2 (i+1)x_{i+1} - \mu_2 ix_i - \mu_3 x_i \tag{18}$$

$$\cdot$$

$$\frac{dx_s}{dt} = \lambda_2 Wx_{s-1} - \lambda_2 Wx_s - \mu_2 sx_s - \mu_3 x_s \tag{19}$$

where

$$\frac{d\sum x_i}{dt} = \frac{dH}{dt} = (\alpha - \mu_3 - \beta H)H - \lambda_2 Wx_s$$

and

$$\frac{d\sum ix_i}{dt} = \frac{dP}{dt} = \lambda_2 WH - \lambda_2 Wx_s(s+1) - \mu_3 P$$

It has not proved possible to solve these equations as yet due to their non linear structure. Some insights into their behaviour, however, can be obtained by consideration of equilibrium states and numerical studies.

Behaviour of the direct life cycle multicompartment model

Lethal level

The size of the lethal level, the parameter s, determines in conjunction with the population rate parameters, the potential carrying capacity of the host population with respect to total parasite population size.

The relative influence of changes in the value of s can be investigated by examining the degree of depression of the host population equilibrium state $\overset{*}{H}$ in the presence of parasitic infection when compared with the steady state achieved by the

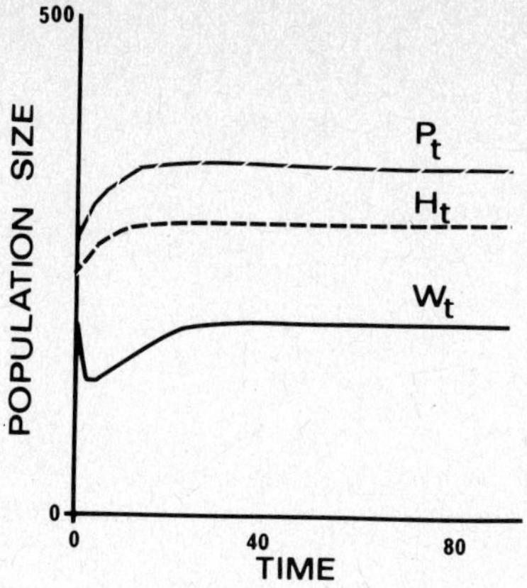

Fig. 5 Dynamics of the single direct life cycle model (equations 13, 14 and 15) regulated by a lethal level mechanism in the absence of other types of density dependent constrains (s=6).

host population $\hat{H}$ in the absence of the parasite. A statistic D can be defined as the depression of the host equilibrium as a result of parasitic infection where

$$D = \frac{\overset{*}{H}}{\hat{H}}$$

The parameter can adopt values between 0 and 1, beeing unity when no depression effect is operating and thus it varies inversely with the degree of depression caused by the parasite. The relationship between degree of host population depression as a result of parasitic infection and s is shown in Fig. 6. The parameter D rapidly moving to a minimum as s increases beyond the value of 10. Alteration of the values of the many rate parameters changes the initial degree of depression but does not influence the general trend of a rapid approach toward a maximum degree of depression at fairly low s values. Changes in lethal burdens between 3 and 10 thus cause a much more marked change in the depression than do changes at much higher values of s.

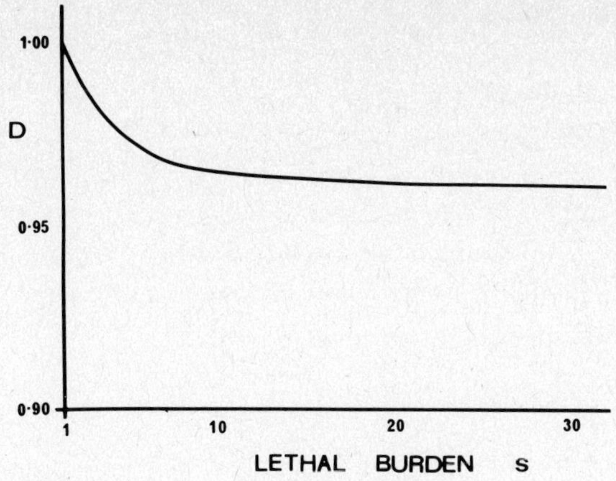

Fig. 6 The relationship between the parameter D and the lethal level s. ($\lambda_1 = 5.0$, $\mu_1 = 2.0$, $\lambda_2 = 0.001$, $\mu_2 = 0.5$, $\alpha = 2.0$, $\beta = 0.01$, $\mu_3 = 0.01$).

An obvious consequence of increasing the lethal burden is for the mean parasite burden per host to increase in a roughly linear fashion with increasing s (Fig. 7).

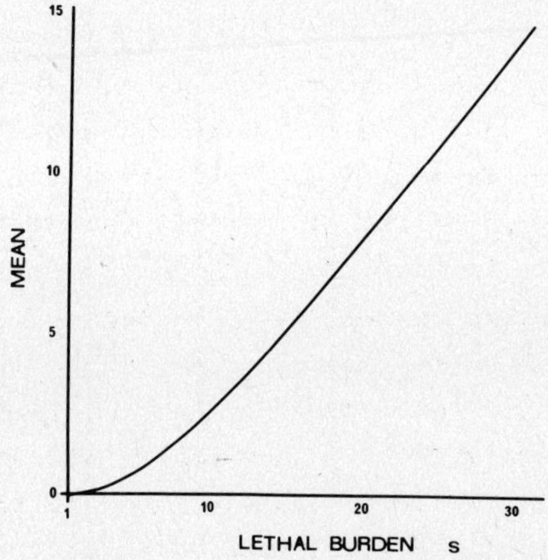

Fig. 7 The relationship between the mean parasite burden per host m where $m = \frac{\sum_i x_i}{\sum x_i}$, and the lethal level s ($\lambda_1 = 5.0$, $\mu_1 = 2.0$, $\lambda_2 = 0.001$, $\mu_2 = 0.5$, $\alpha = 2.0$, $\beta = 0.01$, $\mu_3 = 0.01$).

The increased degree of host population depression resulting from increased values of s is insufficient to cancel out the effects of the increasing mean parasite burden and thus the total parasite population size P also increases with rises in the lethal level.

Gradually increasing the value of s has a marked effect on the growth of each population towards the equilibrium states. At low values the populations quickly achieve a steady state, while high s values lead to a high degree of oscillatory behaviour which in general is damped and thus a steady state is achieve (Fig. 8).

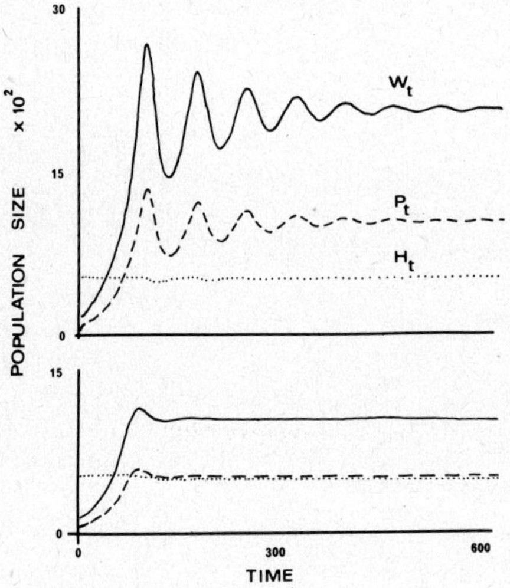

Fig. 8 The trajectories of the three populations involved in the direct life cycle model, W_t $P_t = \sum ix$ and $H_t = \sum x_i$ where (A)s = 12 (B)s = 3.

For certain parameter values, however, the oscillations may not be damped at high s levels, and be either of constant or inc reasing amplitude. The latter situation leads to the extinction of the parasite populations and subsequent recovery of the host population to its equilibrium state $\hat{H}$. This type of behaviour is caused by the abrupt cut off point created by the lethal burden concept. The parasites flow through the system until a large number of hosts reach the lethal burden and are removed from the systems with their parasite loads, causing a sudden decline in all the populations (Fig. 9).

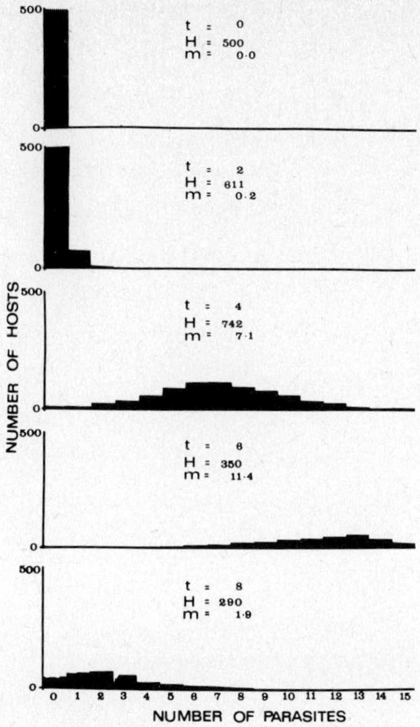

Fig. 9 The time dependency of the frequency distribution of numbers of parasites per host as influenced by the lethal level regulatory mechanism (s = 15, H = host population size, m = mean number of parasites per host, t = time).

This effect is analogous in some respects to the inclusion of time delays or lags in population models, since the regulatory effect in the host parasite interaction only begins to have a marked influence when a sufficient of hosts are approaching the lethal level. Theoretical ecological studies have shown that the inclusion of time lags in the simple logistiv model of population growth leads to oscillations in population size due to the delayed form of regulation (Wangersky and Cunningham, 1956). In general the inclusion of time lags in both single and multispecies models tends to decrease the stability of the system, particularly if the delay is long in comparison to the natural generation times of the organisms involved (May et al, 1974).

The lethal level mechanism acts essentially as a time delayed form of regulation since a lag is involved in the movement of hosts from their birth point and membership of the x_o class to their reaching the x_s class. Both the size of the parameter s and the population rates such as the infection rate λ_2 and the death rate of parasites μ_2 which effect the length of the lag in the delayed form of regulation, determine the degree of oscillatory behaviour exhibited by the equations. Furthermore it is obvious that the initial conditions used in the numerical solution of the equations will have a marked effect on the behaviour of the model, for certain parameter values. For example, the introduction of the infection into an entirely uninfected host population where $H_{t=o} = x_o$, by a positive value

These insights into the influence of the parameter s on the behaviour of the direct life cycle model, have important implications for the stability of host parasite interactions in which a lethal level concept is a realistic biological assumption. It is possible for any value of s to select a set of population rate values which lead to damped oscillations of the populations to steady states. Thus for the survival of a parasite species, on an evolutionary time scale, selective processes may lead to the generation of population rate parameter values, such as egg production and infection rates, which lead to a stable interaction in relation to the lethal burden of parasites required to kill a host species. Conversely it can be argued that for a given set of parameter values, selective processes will operate to generate a particular value of s which leads to regulated growth and stability for the populations involved in a given host parasite burden between strata of the host population, such as age classes, may thus result in the persistance of the association between host and parasite only in certain age groups. For example if the rate parameters are constant the parasite may only persist in young age classes of the host population since increased tolerance to the parasites, and thus a larger s value in older age groups, may lead to the elimination of the parasite due to the increase in the lag in the delayed regulatory mechanisms. Conversely, and perhaps more in line with biological reality, the rate parameters such as infection rates will themselves vary between different strata of the host population, perhaps allowing the continuation of the relationship between host and parasite throughout the entire population.

The abruptness of the lethal level regulatory mechanism is obviously a rather crude representation of the real mechanisms involved in parasite induced host mortality. The mechanism can however be modified in a variety of ways. The parasite death rate, μ_2, may in reality be large, and thus very few hosts will actually acquire a lethal burden of parasites. Alternatively, the parasite death rate may be density dependent and thus a function of the parasite burden i within a single host. More complex models can also be formulated in which the lethal level is assumed to have an expected value $\bar{s}$. The variability of the parameter s within the host population, and thus the frequency distribution of s values will determine the proportions of hosts dying in the various classes. In general, all these mechanisms tend to decrease the abruptness of the regulatory process and thus decrease the degree of oscillatory behaviour exhibited by the model. They do not, however, entirely eliminate the time delay effect created by parasite induced mortalities if these are assumed to depend in some manner on the degree of parasitc infection.

Population rate parameters

The direct life cycle model contains a number of rate parameters influencing the growth and decay of the populations involved and thus the relative influence of each rate on the dynamical properties of the interaction can only be assessed by extensive numerical studies in the absence of analytic solutions.

Certain basic biological features of host parasite interactions allow the elimination of various combination of rate parameters. The parasite egg production rate λ_1 for example will always be many orders of magnitude greater than the host birth rate. Furthermore, the infection rate of the hosts, λ_2, will usually be very small in comparison to parameters such as the birth and death rates of both host and parasite populations.

The relative influences of four particular rates, in conjunction with the parameter s, on the behaviour of the model have been examined in detail. The four rates were selected on the grounds of their potential susceptibility to control measures designed to eradicate or decrease the incidence of direct life cycle parasitic diseases. The statistic, D, measuring inversely the degree of depression of the host population caused by parasitic infection was used to examine the effects of various parameter values.

Death rate of adult parasites (μ_2)

The influence of this parameter on the behaviour of the system is straightforward, increasing values of μ_2 leading to a decrease in the degree of host population depression until ultimately very high death rates cause the extinction of the parasites, (Fig. 10).

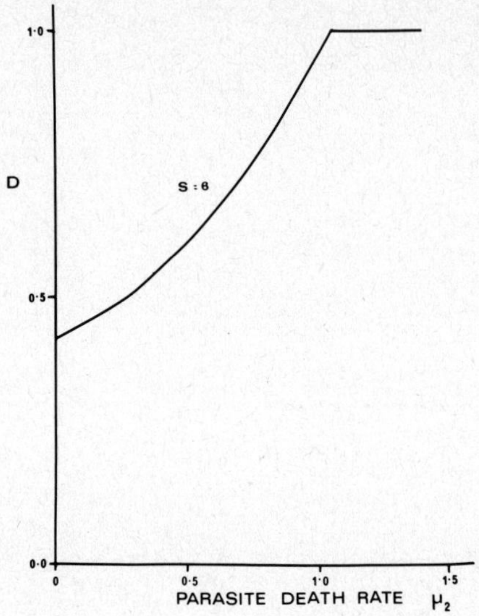

Fig. 10 The relationship between the parameter D and the parasite death rate μ_2 ($s = 6$, $\lambda_1 = 5.0$, $\mu_1 = 2.0$, $\lambda_2 = 0.001$, $\alpha = 2.0$, $\beta = 0.01$, $\lambda_3 = 0.01$).

When the death rate is very slight or zero the value of D is entirely determined by the lethal level the parameter s. In general, increasing values of s do lead to small increases in the degree of depression, although this is less marked when μ_2 is large. Increases in the adult parasite death rate naturally lead to a decrease in the mean parasite burden of the host population (Fig. 11).

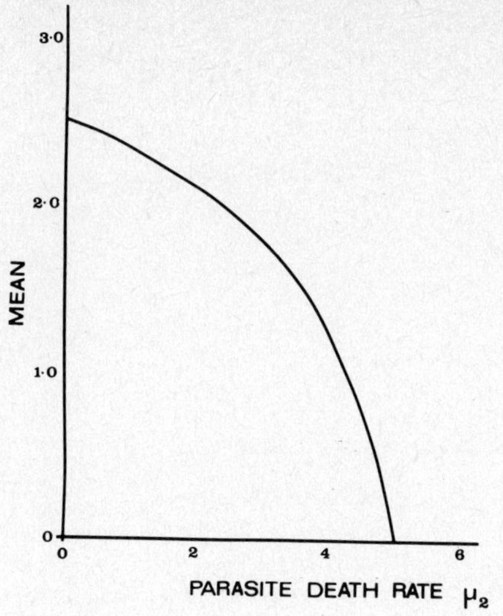

Fig. 11 The relationship between the mean parasite burden m and the adult parasite death rate μ_2 ($s = 6$, $\lambda_1 = 5.0$, $\mu_1 = 2.0$, $\lambda_2 = 0.001$, $\alpha = 2.0$, $\beta = 0.01$, $\lambda_3 = 0.01$).

Death rate of infective stages (μ_1)

When the death rate of the infective stages is zero or nearly so, the pool of infective stages increases at an exponential rate and thus, even if the infection rate λ_2 is very small, the host population tends to become very heavily infected. Ultimately this can lead to the extinction of both the host and parasite populations due to excessive numbers of hosts acquiring lethal burdens of parasites. Conversely when the death rate is very high the parasite population becomes extinct and thus D becomes unity in value. (Fig. 12). Increases in this death rate again lead to a decrease in the mean parasite burden but also a decrease in the total number of parasites present in the habitat occurs even though the host population is larger due to high values of D.

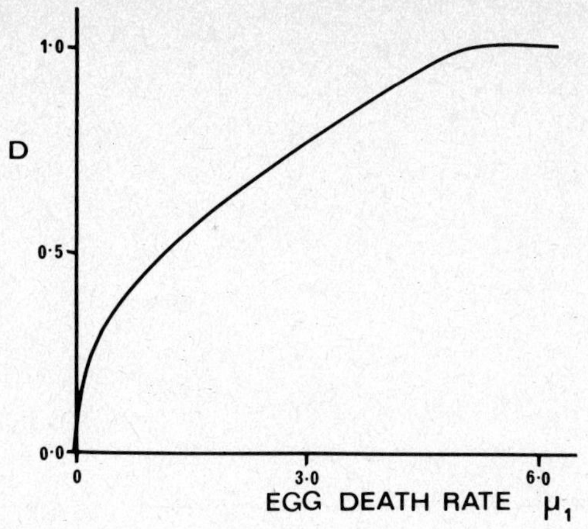

Fig. 12 The relationship between the parameter D and the parasite egg death rate μ_1 ($s = 6$, $\lambda_1 = 5.0$, $\lambda_2 = 0.001$, $\mu_2 = 0.5$, $\alpha = 2.0$, $\beta = 0.01$, $\lambda_3 = 0.01$).

Birth rate of infective stages or eggs (λ_1)

Very low values of λ_1, the precise level depending on other parameter values such as μ_2, can be insufficient to sustain parasite populations and thus extinction can occur. A type of threshold effect is generated, however, where rate values above a certain level lead to a marked degree of depression of the host population (Fig. 13). These low levels of D, with increasing egg production, eventually plateau out and thus marked increases in parasite reproduction do not have any further detrimental effects on the host population equilibrium level. The mean parasite burden rises with increasing values of λ_1 until a plateau is again reached (Fig. 14).

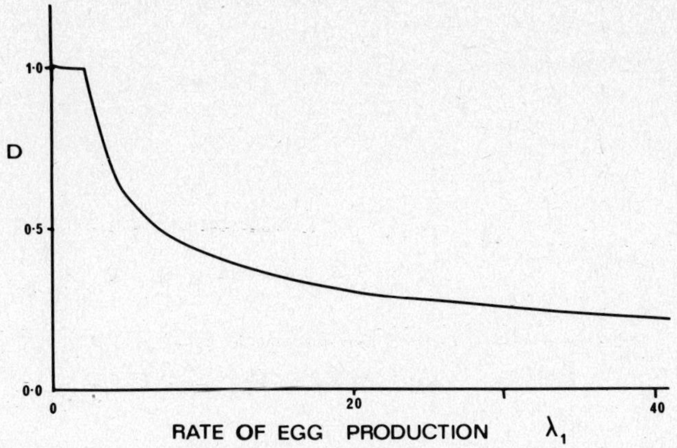

Fig. 13 The relationship between the parameter D and the parasite egg production rate λ_1 ($s = 6$, $\mu_1 = 2.0$, $\lambda_2 = 0.001$, $\mu_2 = 0.5$, $\alpha = 2.0$, $\beta = 0.01$, $\lambda_3 = 0.01$).

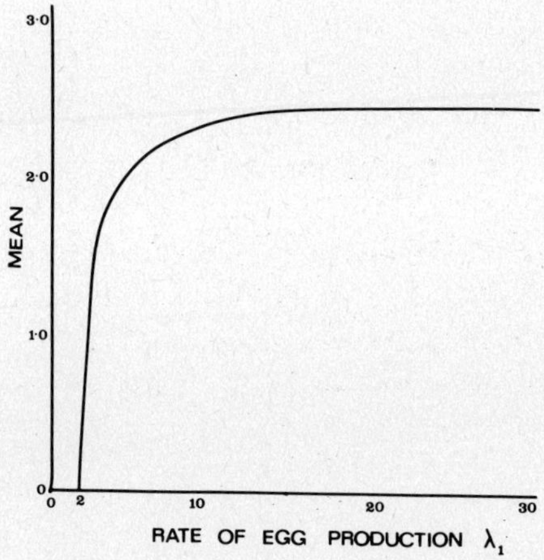

Fig. 14 The relationship between the mean parasite burden m and the parasite egg production rate λ_1 ($s = 6$, $\mu_1 = 2.0$, $\lambda_2 = 0.01$, $\mu_2 = 0.5$, $\alpha = 2.0$, $\beta = 0.01$, $\lambda_3 = 0.01$).

Interestingly, these results suggest that there may be an optimum reproductive rate for a given parasite species above which the returns gained for increased investment of energy in egg production are of little value in terms of increased population size.

Rate of infection of hosts (λ_2)

Small alterations in the value of this particular parameter cause very pronounced changes in the behaviour of the model. If the rate is very low, then naturally the parasite fails to establish in the host population. Above the critical level at which the parasite gains a foothold (a level determined by other rate values such as λ_1 and μ_2) small changes in infection rates lead to considerable changes in the degree of host population depression (Fig. 15).

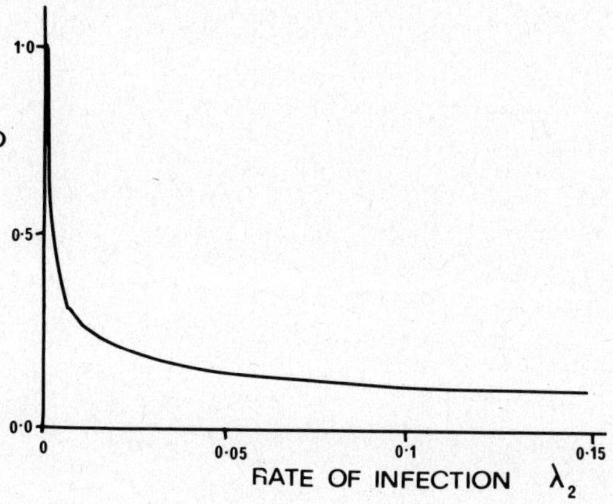

Fig. 15 The relationship between the parameter D and the host infection rate λ_2 ($s = 6$, $\lambda_1 = 5.0$, $\mu_1 = 2.0$, $\mu_2 = 0.5$, $\alpha = 2.0$, $\beta = 0.01$, $\lambda_3 = 0.01$).

The degree of depression, however, eventually plateaus out as λ_2 becomes large with the host population subsisting, at a very low level due to heavy parasite burdens. The mean parasite burden similarly increases to a plateau with increasing infection rates, and thus small changes at high infection rate levels because little difference in parasite load per host (Fig. 16).

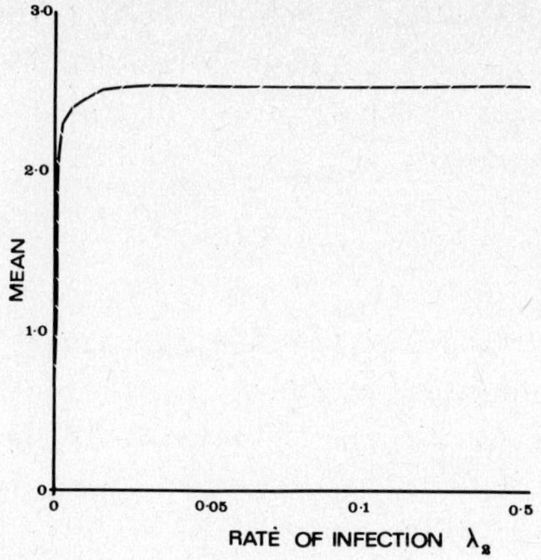

Fig. 16 The relationship between the mean parasite burden m and the host infection rate λ_2 (s = 6, λ_1 = 5.0, μ_1 = 2.0, μ_2 = 0.5, α = 2.0, β = 0.01, λ_3 = 0.01).

This pattern generates an interesting population result, since increases in λ_2 lead to both an increased degree of depression of the host population and an increased parasite burden, but the changes in the latter statistic are insufficient to prevent a decrease of the total parasite population size. Increased efficiency in the rate of host infection may thus be counter productive in terms of the total number of parasites harboured by a host population.

The behaviour of the model in relation to changes in the values of both λ_1 and λ_2, suggests that these particular rates may be of importance in the design of parasite control programs. For example, measures which result in very small negative changes in the rate of infection may lead to considerable degree of relief to the host population in terms of decreasing the amount of depression caused by parasitic infection. If the infection rate is large, however, small or large changes engineered by control measures may not lead to any great deal of relief to the depression of the host population (Fig. 14). Chemotherapeutic agents which suppress parasite reproductive rates (λ_1) can also have very marked beneficial effects on the host population. It is interesting to note in connection with this particular rate, that in many host parasite interactions, the hosts immune responses often result in a suppression of parasite reproductive potential as shown for example in the cases of Trypanosoma musculi infection in mice (Viens et al, 1975) and infections of the nematode Ostertagia ostertagi in calves (Michel, 1967).

THEORY VERSUS REALITY

The comparison of theory and reality naturally necessitates the acquisition of accurate estimates of the many rate parameters involved in parasite life cycle models. Unfortunately, however, very few field or laboratory studies have yielded such quantitive estimates of the population processes such as infection rates, rate of host mortality caused by parasite infection or even survival rates of the various parasitic stages in the life cycle. The majority of quantitive work has been concerned with deseases of economic importance such as malaria (Macdonald, 1957; Dietz et al, 1974) and schistosomiasis (Cohen, 1973 a). In general, however, it is surprising how little is known concerning the rate parameters controlling the dynamics of host- parasite interactions.

The paucity of our knowledge is no doubt due to the complexities of parasite life cycles, but the lack of experimental and field information is also a direct result of the intimacy of the relationship between host and parasite. For instance an observational task such as counting of animal numbers, which is comparatively easy in the case of free living organisms, generates many practical problems when considering endoparasitic species. In the investigation of parasitic diseases of man, destructive sampling is obviously not possible and information on parasite burdens can often only be obtained from post mortem examinations.

Theoretical models can play an important role in the collection of data, since not only do they precisely define the rates requiring estimation, but often they pinpoint particular parameters of importance which contribute disproportionally to the behaviour of the model system. This latter point is clearly illustrated by the influence of the egg production rate λ_1, and the infection rate λ_2 in the direct life cycle model.

Much work remains to be done on the adequacy of such models, however, since they contain many simplifying assumptions. In particular, many of the rate parameters involved in a given life cycle, will not be constants but may be functionally dependent on population density, due either to intraspecific competition or to immunological attack, or on environmental parameters such as temperature. Density independent effects will be particularly important in the case of free living stages in parasite life cycles and for parasitic stages within poikilothermic hosts such as the mosquito or molluscan hosts (Fig. 17).

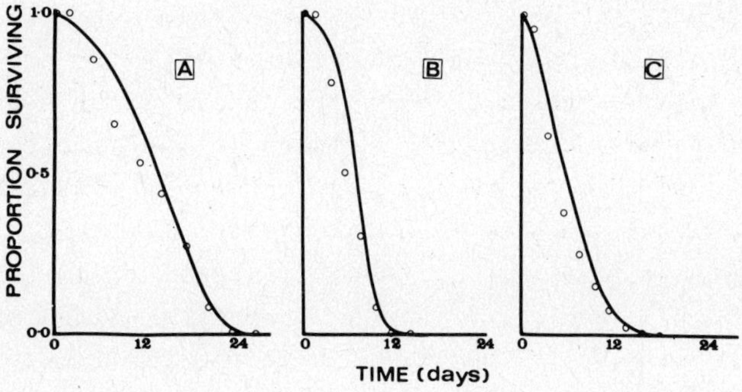

Fig. 17 The influence of temperature on the survival characteristics of populations of the larval stage of Bunostomum trignocephalum (data from Narain, 1965). Graph A 10° C Graph B 0° C Graph C 25° C.
Open circles - observed data. Solid line - survival curves predicted by an age and temperature dependent death model of the form

$$\frac{dW_t}{dt} = -\mu(T, t)\, W_t.$$

Rate parameters may also be age dependent such as the survival of larval or adult parasites, or the rate of egg production in adult helminths. Fig. 18, for example, illustrates the age dependency in survival and egg production of an ectoparasitic digenean Transversotrema patialensis which has a life cycle very similar to the schistosome parasites.

Many problems in mathematical description are raised by immunological responses, since these mechanisms are invariably complex in nature. It is becoming increasingly apparent that the majority of interactions, whether involving protozoon or helmith parasites within either vertebrate or invertebrate hosts, are to some extend characterised by the presence of host generated immune responses elicited by the presence of the parasite. These responses may act to decrease the rate of establishment of invading infective stages (Jenkins and Phillipson, 1971), increase the death rate of established larval ar adult parasites (Mulligan et al, 1965; Jarrett et al, 1968) and either decrease the rate of egg production by helminth parasites (Michel, 1967) or inhibit the reproductive rate of protozoon parasites which multiply within the host (Viens et al, 1974; Bradley, 1971). In practice, one or more population parameters may be influenced by the hosts' defence mechanisms.

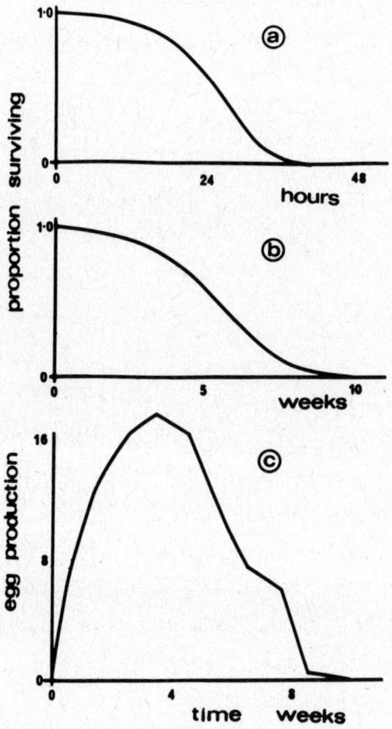

Fig. 18 Age dependent survival and egg production of the ectoparasitic digenean at 24° C. <u>Transversotrema patialense</u> <u>Graph A</u> Cercarial survival (after Anderson and Whitfield, 1975) <u>Graph B</u> Adult parasite survival on the host (Anderson et al, 1976). <u>Graph C</u> Egg production per fluke (C. Mills, Kings's College, London University, unpublished data).

Selective action may occur in certain relationships either on specific age groups of parasites as suggested by the work of Ormerod (1963) on the survival of Trypanosoma lewisi in the rat, or as a result of the age of the host as demonstrated by Phillips et al (1975) in the case of infections of Schistosoma mansoni also in the rat. The functional relationship between the immune response of the host and the parasites population rate parameters may broadly speaking, be either time or density dependent or both. In this context time dependency implies that the severity of the immune response is some function of the time from the initial parasitic invasion. The period elapsing from initial entry of a parasitic infection to a detectable antibody response is often termed the induction period and this lag is generally shorter on subsequent exposure to infection.

Some parasitic species are capable of avoiding the hosts' immunological attack by changing the antigenic nature of their surface coating. Such a response is shown by Trypanosoma brucei rhodesiense in the rat (Gray, 1962) and leads to cyclic oscillation of parasite population size within a single host due to continual interplay between the hosts' immune response and the parasite's surface structure. Helminth parasites may also avoid immuniological attack by encorporating host antigen into their surface coating. This mechanism is adopted by Schistosoma mansoni in the rhesus monkey, the parasites disguising itself from the humoral antibody response of the host (Smithers et al, 1969; Hockley and McLaren, 1972).

Genetic factors are also of considerable importance in determining the effectiveness of an immune attack as recently shown by Wakelin (1975), who compared survival of the nematode Trichuris muris in random and inbred strains of mice and found marked differences between the two groups of hosts.

An added complication in trying to quantify the nature of an immune mechanism arises in certain parasitic diseases, where initial infections of one species of parasite may confer some degree of resistance in the host to subsequent infections with different species of parasite (Cohen, 1973 b). Alternatively, parasite survival may be inhanced by previous infection with another species, due to immunodepression resulting in inhibition of the host's responses on subsequent infections with a new species (Cox, 1975).

The importance of time delays in the dynamics of complex population interactions has already been demonstrated in this paper, in connection with the influence of altering the lethal parasite burden in the direct life cycle model. Many other biological processes, however, give rise to time delays, such as development lags which occur for example between establishment of a helminth parasite within a host, and the beginning of egg production. Similar delays in protozoon parasites such as species of Plasmodium where a development lag occurs from the ingestion of gametocytes via a blood meal to the production of sporozoites and transfer of the parasite to a new host. These delay may in many cases be dependent on environmental factors such as temperature, particularly in the case of parasites within poikilothermic hosts or free living infective stages (Fig. 19).

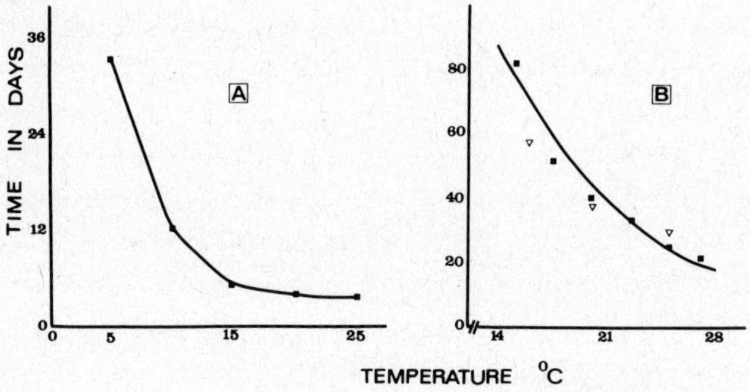

Fig. 19 The influence of temperature on parasite development rates. Graph A Effects of temperature on the time taken by Dictyocaulus viviparous larvae to develop to the L_3 stage (data from Rose, 1956). Graph B Effects of temperature on the time taken by rediae of Fasciola hepatica in Lymnaea truncatula to develop to their maximum weight and begin production of cercariae. (Open triangles - data from Ollerenshaw, 1971; solid squares - data from Nice and Wilson, 1974).

When faced with such complexity, the formulation of realistic models appears difficult and published models seem very crude and simplistic. Attempts to construct models of the dynamics of the many populations involved in protozoon and helminth life cycles are invariably deterministic in nature. Stochastic formulations are undoubtedly ideally required, particularly when the behaviour of stochastic and deterministic models of the simple epidemic are compared (Bailey, 1957). It is possible to write down a stochastic model, consisting of a series of partial differential equations for the various generating functions of the numerous populations in the direct life cycle system (Fig. 1), but these are of necessity, highly non linear in nature and thus analytical results are difficult to obtain at present. Stochastic models can, however, in certain circumstances be formulated to describe the dynamics of a population within a specific compartment of the life cycle, such as the immigration- death process controlling the adult parasite population size within a single host. This approach can be useful in providing estimates for specific parameters as illustrated by Dietz et al (1974) in the model of malaria dynamics in the African savannah.

Deterministic models for all their shortcoming, do provide a basis for later elaboration and can lead to conceptual insights into host parasite interactions. Furthermore, stability analysis of equilibrium states of the populations involved in such models may be helpful in suggesting possible epidemiological control mechanisms. The analysis will ideally provide information concerning the areas of parameter values in the total parameter space which lead to either stable or unstable equilibrium states. Control measures designed to eradicate or reduce the incidence of a parasitic disease should perhaps aim to alter a rate parameter, such as the rate of infection, or death rate of adult parasites, such that it moves into the area of unstable parameter space, where small perturbations of the system will result in the extinction of the parasite species in a given host population.

An additional approach worth considering in attempts to understand the dynamics of host parasite interaction is the use of laboratory biological models. Various species of both helminth and protozoon parasites with direct and indirect life cycles can be maintained in the laboratory and thus provide experimental models for the examination of various population processes and regulatory mechanisms. This approach has been adopted by Anderson and Whitfield (1975) utilising an ectoparasitic digenean <u>Transversotrema patialense</u> with an indirect life cycle of similar structure to the schistosome parasites. This work aims to provide quantitive estimates of the many parameters involved in the indirect life cycle and thus to provide a means of examining experimentally the dynamic behaviour of the system. Due to the scarcity of detailed experimental information concerning the ecology of the majority of eucaryotic parasites, it would seem worthwhile searching for other experimental models with life cycles resembling parasites of economic importance.

With the acceleration of interest in mathematical models of host parasite interactions, perhaps the most overiding necessity is for the acquisition of good experimental and field data. The collection of such data should ideally proceed hand in hand with theoretical studies on the dynamic behaviour of the population interactions. Such an approach should ensure a two way interaction between theoretical models and biological reality, and thus greatly aid our understanding of the epidemiology of protozoan and helminth parasitic diseases.

REFERENCES

Anderson, R. M. (1974): "Mathematical models of the host - helminth parasite interactions" Ecological stability (Ed. by M. B. Usher and M. H. Williamson) pp 43 - 69 Chapman and Hall, London.

Anderson, R. M. and Whitfield, P. J.(1975): Survival characteristics of the free- living cercarial population of the ectoparasitic digenean Transversotrema patialensis. (Soparker, 1924). Parasitology 70, 295 - 310.

Anderson, R. M., Whitfield, P. J. and Mills, C. A. (1976): Experimental population dynamics of the ectoparasitic digenean Transversotrema patialense on the fish host Brachydanio rerio. In preparation.

Bailey, N. T. J. (1957): The mathemati cal theory of epidemics Griffin, London.

Boray, J. C. (1969): Experimental fascioliasis in Australia. Advances in Parasitology, 7, 96 - 210.

Bradley, D. J. (1971): Inhibition of Leishmanis donovani reproduction during chronic infections in mice. Trans. R. Soc. Trop. Med. Hyg. 65, 17 - 18.

Cohen, J. E. (1973 a): Selective host mortality in a catalytic model applied to Schistosomiasis, Am. Natur., 107, 199 - 212.

Cohen, J. E. (1973 b): Heterologous immunity in human malaria. Quant. Rev. Biol. 48, 467 - 489.

Cox, F. E. G. (1975): Enhanced Trypanosoma musculi infections in mice with concomitant malaria. Nature, 258, 148 - 149.

Crofton, H. D. (1971 a): A quantitive approach to parasitism. Parasitology, 62, 179 - 193.

Crofton, H. D. (1971 b): A model of host - parasite relationships. *Parasitology*, 63, 343 - 364.

Dietz, K., (1967): Epidemics and rumours: A survey. *J. Roy. Statist. Soc. Ser. A*, 130, 505 - 528.

Dietz, K., Molineaux, L. and Thomas, A. (1974): A malaria model tested in the African savannah. *Bull. Wld. Hlth. Org.* 50, 347 - 357.

Goffman, W. and Warren, K. S. (1970): An application of the Kermack - McKendrick theory to the epidemiology of Schistosomiasis. *Am. J. Trop. Med. Hyg.* 19, 278 - 283.

Gray, A. R. (1962): The influence of antibody on serological variation in *Trypanosoma brucei*. *Ann. Trop. Med. Parasit.*56, 4 - 13.

Greenwood, M. and Yule, G. U. (1920): An inquiry into the nature of frequency distributions representative of multiple happenings with particular reference to the occurence of multiple attacks of disease or of repeated accidents. *J. Roy. Stat. Soc.* 83, 255 -

Hockley, D. J. and McLaren, D. J. (1972): *Schistosoma mansoni*: changes in the outer membrane of the tegument during development from cercariae to adult worm. *Int. J. Parasit*. 3, 13 -

Jarrett, E. E. E., Jarrett, W. F. H. and Urquhart, G. M. (1968): Quantitative studies on the kinetics of establishment and expulsion of intestinal nematode populations in susceptible and immune hosts, *Nippostrongylus brasiliensis* in the rat. Parasitology, 58, 625 - 639.

Jenkins, D. C. and Phillipson, R. F. (1971): The kinetics of repeated low - level infections of *Nippostrongylus brasiliensis* in the rat. *Parasitology*, 62, 457 - 466.

Kostitzin, V. A. (1934): *Symbiose, parasitisme et evolution*. Hermann, Paris.

Leslie, P. H. and Gower, J. C. (1960): The properties of a stochastic model for the predator - prey type interaction between two species. *Biometrika,* 47, 219 - 234

Leyton, M. K. (1968): Stochastic models in populations of helminthic parasites in the definitive host. 11 - Sexual mating functions. Math. Biosci. 3, 413 - 9.

Lotka, A. J. (1925): Elements of physical Biology. Williams and Wilkins, Baltimore.

May, R. M. (1973): Stability and Complexity in Model Ecosystems. Princeton University Press, Princeton.

May, R. M., Conway, G. R., Hassell, M. P. and Southwood, T. R. E. (1974): Time delays, density dependence and single species oscillations. J. Anim. Ecol. 47, 907 - 15.

Mcdonald, G. (1950): The analysis of infection rates in diseases in which superinfection occurs. Trop. Dis. Bull. 47, 907 - 915.

Mcdonald, G. (1957): The epidemiology and control of malaria Oxford University Press, London.

McKendrick, A. G. (1926): Application of mathematics to medical problems. Proc. Edinb. Math. Soc. 44, 98 - 130.

Michel, J. F. (1967): Regulation of egg output of populations of Ostertagia ostertagi. Nature, 215, 1001 - 1002.

Mulligan, W., Urquhart, G. M., Jennings, F. W. and Neilson, J. T. M. (1965): Immuno ogical studies of Nippostrongylus brasiliensis infections in the rat, the self- cure phenomenon. Expt. Parasit. 16, 341 -

Narain, B. (1965): Survival of the first stage larvae and infective larvae of Bunostomum trgnocephalum Rudolphi, 1808. Parasitology, 55, 551 - 558.

Nassell, I. and Hirsch, W. M. (1972): A mathemati al model of some helminthic infections. Com. Pure Appl. Math. 25, 459 - 77.

Nassell, I. and Hirsch, W. M. (1973): The transmission dynamics of Schistosomiasis. Com. Pure Appl. Math. 26, 359 - 453.

Nice, N. G. and Wilson, R. A. (1974): A study of the effect of temperature on the growth of Fasciola hepatica in Lymnaea truncatula Parasitology, 68, 47 - 56.

Ollerenshaw, C. B. (1971): Forecasting liver fluke disease in England and Wales 1958 - 1968, with a comment on the influence of climate on the incidence of the disease in some other countries. Vet. Med. Rev. 213, 289 - 312.

Omerod, W. E. (1973): The initial stages of infection with Trypanosoma lewisi: control of parasitaemia by the host. Immunity to protozoa (Eds. P. C. C. Garnham, A. E. Pierce and I. Roitt). pp 213 - Blackwell Scientific publications, Oxford.

Pan, C. T. (1965): Studies on the host parasite relationship between Schistosoma mansoni and the snail Australorbis glabratus. Am. J. Trop. Med. Hyg. 14, 931 - 976.

Pearl, R. and Read, L. J. (1920): On the rate of growth of the population of the United States since 1790 and its mathematical representation. Proc. Nat. Acad. Sci. U. S. A. 6, 273 - 288.

Phillips, S. M., Reid, W. A., Bruce, J. I., Hedlund, K., Colvin, R. C., Campbell, R., Diggs, C. L. and Sedun, E. H. (1975): The cellular and humoral immune response to Schistosoma mansoni infections in inbred rats. 1 Mechanisms during initial exposure. Cell Immun.19, 99 - 116.

Pielou, E. C. (1969): An Introduction to Mathematical Ecology, Wiley Interscience, New York.

Rose, J. H. (1956): The bionomics of the free living larvae of Dictyocaulus viviparous. J. Comp. Path. Ther. 66, 228 - 240.

Smithers, S. R., Terry, R. J. and Hockley, D. J. (1969): Host antigens in Schistosomiasis. Proc. Roy. Soc. 171, 483 -

Tallis, G. M. and Leyton, M. (1969): Stochastic models of populations of helminthic parasites in the definitive host - 1 Math. Biosci. 4, 39 - 48.

Viens, P., Targett, G. A. T., Leuchars, E. and Davies, A. J. S. (1974): The immunological response of mice to Trypanosoma musculi 1. Initial control of the infection and the effect of T- cell deprivation. Clin. Exp. Immunol. 16, 279 - 294.

Wakelin, D. (1975): Genetic control of immune responses to parasitic infections of Trichuris muris in inbred and random bred strains of mice. Parasitology, 71, 51 - 60.

Wangersky, P. J. and Cunningham, W. J. (1956): On time lags in equations of growth. Proc. nat. Acad. Sci. U.S.A. 42, 699 - 702.

ABSTRACT MODEL AND EPIDEMIOLOGICAL REALITY OF INFLUENZA A

A. Fortmann
Lüneburg

Every theory of infection spread may be considered as a simplified model of the course of the real epidemic. A mathematical model is a simplified image describing the progress of an epidemic using a small number of epidemiological parameters. By changing these parameters the model puts out and simulates different outcomes of the epidemics. The theoretical calculations lead to results which can be used to discuss unexplainable phenomena of an epidemic.

I was lead to this influenza model by observations made during the influenza epidemics of 1968 to 1970 and by a chain letter which I had to copy five times for my daughter.

A practicing physician can easly observe the progress of an influenza epidemic by looking at the number of the daily consultations about "infection with temperature". The incidence and prevalence of cases reported to a regional health agency plotted versus time shows a one week shift between the two parameters of the epidemic. This does not make much difference for very retarded epidemics. If we look at the A/Hong Kong epidemic of December and January 1969/70, however, which had a very rapid spread and peaked already three weeks after its onset we can see that the prognosis by public health agencies lead to grotesque miscalculations.

Figure 1 shows as epidemics parameters the monthly virus verifications from 21 virological laboratories within West-Germany during the two A/Hongkong epidemics of 1968/69 and 1969/70:

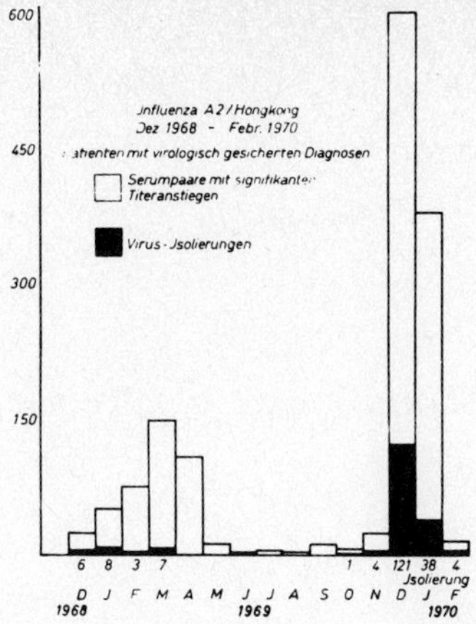

Fig. 1 Virus verification during the influenza A epidemic 1968/1970

The first epidemic lasted approximately five months whereas the second one lasted less than 2 months. Virological checks proved that during the two months approximately one third of the population was infected by influenza virus, during the previous 5 months epidemic, however, only one fifths of the population.

The analysis of the progression of the two epidemics led to two results:

1. Incidence and prevalence show a constant phaseshift of one week independent of the progression of the epidemic. Only the incidence data are useful for making an estimate of the status of the epidemic.
2. Duration of the epidemic and percentage of infected and later immune persons are reciprocal to each other.

The recreational contemplation of the problems of chain letters created the idea that the chain reaction of influenza infections and the spread of the disease may follow the same principle. The rules of the game are as follows: An infectious patient infects several susceptible persons during the period of virus excretion. After the latency period they in turn infect a new number of susceptibles and so on. The statistical mean of contacts between persons during the excretion period is called the contact number q.

Density and mobility of the population are important parameters determining the contact number. As an example, we may expect extremely large numbers in a youth camp, on the other hand, low numbers may be expected in a thinly populated desert area.

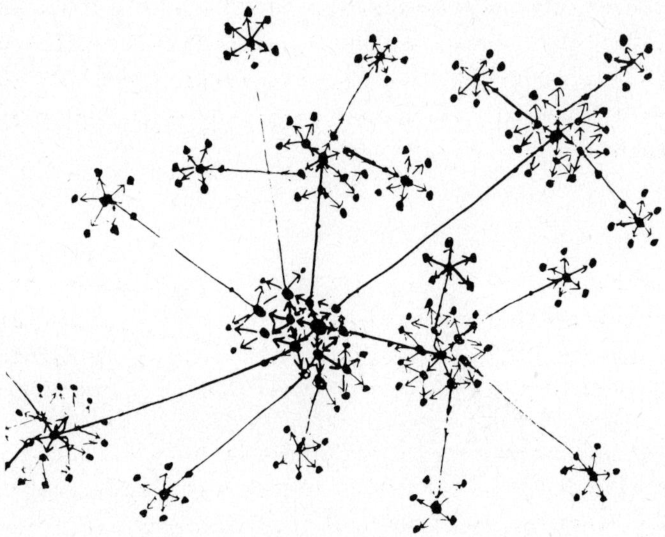

Fig. 2 Spatial propagation of an epidemic using contact number q = 6

Figure 2 shows the spread of an infection with the contact number of q = 6. The population is assumed to be fully susceptible i.e. every infected person will develop symptoms of the disease. The first case of the disease is shown at the center. He infects 6 persons in his environment, two of them remain in his proximity, 4 persons move away and spread the disease there; for instance, at their place of work. In the third generation again some spread the disease in their immediate surroundings while others move away and spread the disease there.

The plot ends with the disease in the fourth generation. In a fully mobile, infinitely large population there should be 6 cases in the second generation, 36 cases in the third, and 216 cases in the fourth generation. Actually in our example there are only 32 cases in the third generation and 132 cases in the fourth. This is a saturation phenomenon showing that the geometrical series type of spread of the infection is limited by the fact that in a population limited by number and space (S) infectious persons keep meeting persons which are already infected or have lived through the disease: We call them immune. The number of immune persons increases just as the number of susceptibles or nonimmunes (N) decreases. The probability that an infectious person meets a susceptible one is approximately equal to the quotient N/S. For the coefficient (W) indicating the number of persons which we expect to be infected by carriers of the infection we have

$$W = q \cdot N/S \tag{1}$$

The number W which we call coefficient of spreading is larger than unity at the start of the experiment. As the number of susceptibles decreases some of the infection chains stop because the infectious person meets more and more immune persons. This process stops the epidemics.

This was the basic conception of the influenza model assuming that all persons of the group are susceptible and that all people develop symptoms of the disease. The assumption of a fully susceptible population is only true for remote areas (for instance Iceland during WW I). Here in West Germany the proportion of persons showing manifest symptoms is 10-15 percent of the total population in large influenza epidemics, 5-10 percent in medium epidemics, and 2-5 percent in small epidemics. The percentage of children is usually larger than that of senior citizens. The true proportion of infected persons is considerably larger, though, because of the high proportion of clinically inapparent infections as can be shown by serological screening. Past epidemics always leave partial immunity in a population even to new variants of the virus. The proportion of immune persons within a population is called the basic immunity I_o in the influenza model including specific and nonspecific protection factors.

In our model, at the start of an epidemic or time t=0 the population S is divided into the group of immunes I and susceptibles N. Because of the small case fatality rate the population can be assumed as being constant

$$N_o + I_o = S \tag{2}$$

During the epidemic a third group, infectious persons K, must be added and we have for any time t:

$$K_t + N_t + I_t = S \tag{3}$$

The number of susceptibles at any time t is:

$$N_t = S - (I_t + K_t) \tag{4}$$

The duration of a generation or generation period of influenza is the period from the infection to the start of excretion of virus by the infectious persons. It is defined as the mean time between infection of a person and spread of the infection to other persons. The generation period is an essential parameter and determines the duration of an epidemic. Measles have a generation period of 10-15 days, mumps 16-25 days, and hepatitis A 5-6 weeks.
That is why we have such long epidemics of measles, mumps, and hepatitis A. It is an accepted result of empirical epidemics research that the duration of an epidemic is a function of the generation period.
Influenza A has an extremely short generation period of 2-3 days - in our influenza model 2 days exactly were assumed. The introduction of a mean generation period makes it possible to abstracize the time of infection in a natural infectious chain which is distributed randomly in time into a simultaneous infection time. In other words: Alls infections occur simultaneously at the same time at the beginning and again, in the next generation, at the end of the period.

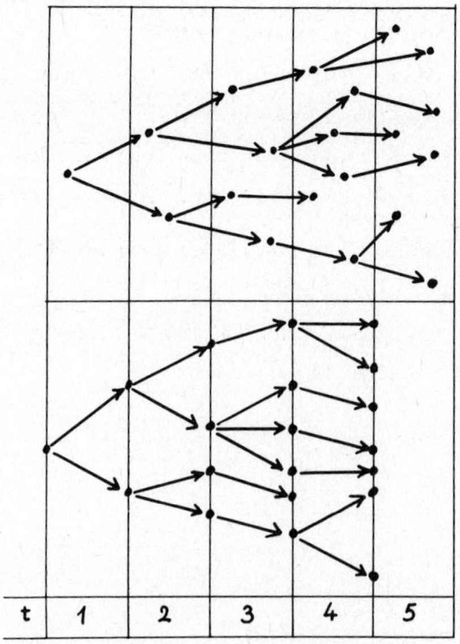

Fig. 3 Top: Infection chains with randomly occurring infection time

Bottom: Synchromized infection time used in the model

The relationship between the number of infectious persons K_t during period t to the expected incidence of the disease during (t + 1) is as follows:

$$K_{t+1} = K_t \cdot W_t = \frac{q}{S} \cdot K_t \cdot N_t \qquad (5)$$

N_t must be calculated for every generation

$$N_t = S - (I_t + K_t) \qquad (4)$$

The number of immunes I_t is the basic immunity I_o plus the sum of all diseased persons up to time t which are considered immune (by definition) after living through the infectious phase of the disease.

$$I_t = I_o + \sum_{t=1}^{t-1} K_t : \qquad (6)$$

Equation 4 must be modified because persons having the disease at time t are no longer susceptible and must be added to the immunes:

$$N_t = S - (I \cdot + \sum_{t=1}^{t} K_{t'}) \qquad (7)$$

Influenza A epidemics spread with a velocity of 100-150 km per day. An epidemic can spread over Central Europe within 2 weeks. After the initial importation of the infection into a large metropolitan area further import of infectious persons can be neglected since import and export of infectious persons add up to zero. This is because between large communities there is always a balance between persons travelling to and from and since the prevalence of the disease of neighbouring cities is approximately equal because of the high velocity of spread.

The following simulations show the progress of a large and a medium influenza A epidemic. The contact number was chosen to be 5, population S = one million and the epidemic was started with 100 infectious persons (K_o = 100).

Simulation No. 1

Basic immunity I_o is 73 percent, susceptibles therefore 27 percent. Incidence K, immunity of the population I, and the spreading factor W are plotted against time in figure 4.

The plot of incidence first shows an exponential rise which soon levels off and after culmination falls approximately like the mirror image of the rise. The fall is slightly steeper than the rise. The right shift of the maximum is typical for retarded epidemics. The maximum of the simulated epidemic occurs on the fourtieth day after the start. The incidence is approximately 5500 persons per day, duration is 75 days. The total incidence is 13,2 percent of the population or 132000 persons which become ill during the epidemic (= 49 % of the susceptibles). The progress of parameter I demonstrates the rising immunity of the population.

Fig. 4

Number of immunes, spreading coefficient, and incidence of a simulated epidemic

Deviation from unity (Abweichung)

Threshold of immunity (Grenzwert)

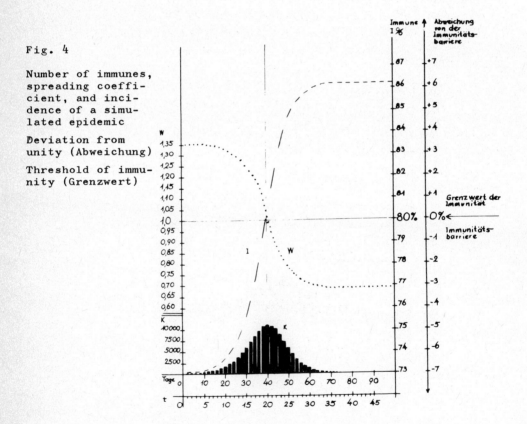

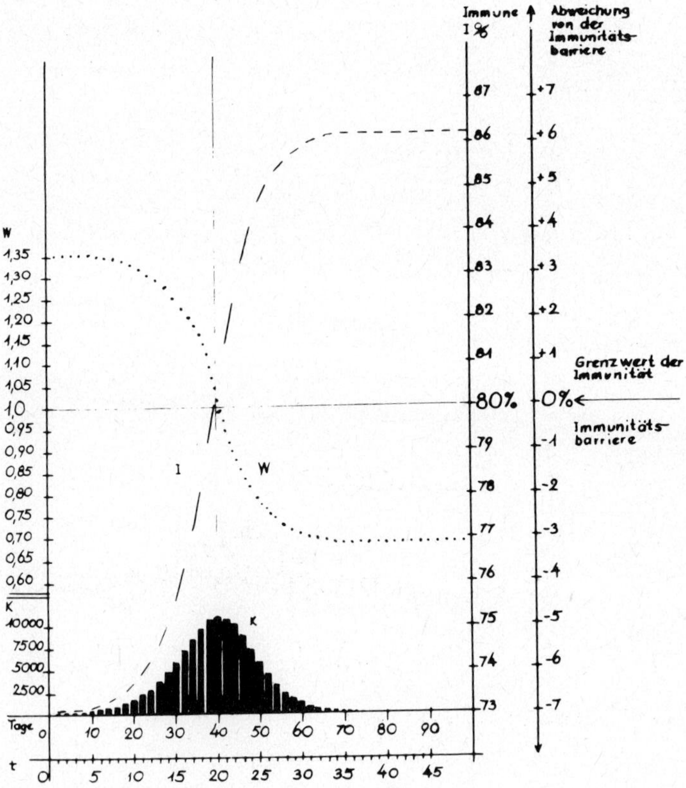

I is equal to the area under curve K swept out at time t since every diseased person becomes immune against a new infection.

As the population gets more and more immune the spreading coefficient decreases. At the maximum of the epicemic the spreading coefficient is equal to unity and the immunity 80 percent. This is called the immunity barrier which will be defined later. As immunity reaches 80 percent and ata spreading coefficient smaller than unity a smaller and smaller number of infectious persons meets an increasing number of immunes. The epidemics thus gets choked of by itself without all susceptibles beeing infected.

Simulation No. 2

Figure 5 shows the second simulation with a basic immunity of I_o = 75 percent. This leads to a considerably flatter progression of the epidemic, the maximum has shifted to the right and occurs on the 50th day. Duration of the epidemic is 93 days, 25 percent of the population or 38 percent of the susceptibles become ill.

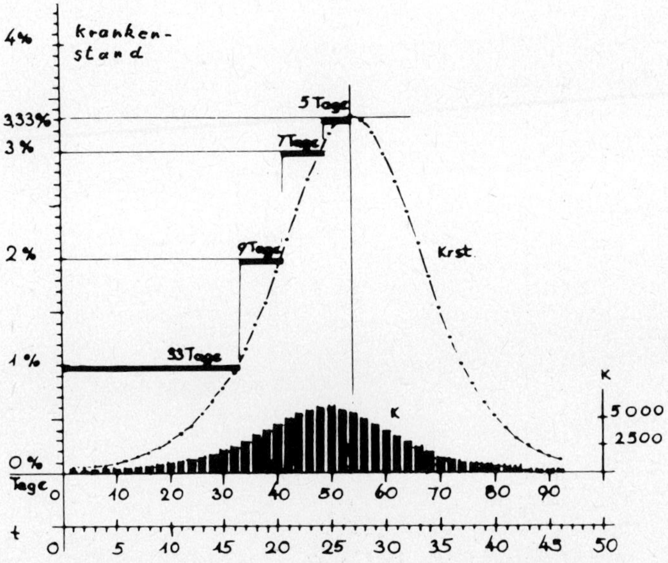

Fig. 5 Prevalence and number of infectious persons
Simulation No. 2

The plot of prevalence as the sum of all simultanously ill persons shows the previously mentioned time shift between incidence and prevalence. The time shift is equal to one half of the mean "time off work" period specific for the disease causing the epidemic.
In the model it is 6 out of 12 days assumed as the "time off work" period. The epidemics discussed earlier in this paper showed a time shift of 7 days disclosing a mean "off work" period of 14 days.
A rise of 1 percent in the number of persons "off work" takes 33 days. Experience shows that this is the time when the epidemic is first noticed. The prevalence and incidence curves first have a very slow rise explaining the phenomenon of a period up to 6 weeks between a first verification of the virus and the outbreak of a manifest epidemic (WOHLRAB).
The next percent of prevalence takes only 9 and another percent 7 days. Now the spread of the epidemic is called "explosive" which is incorrect at least in an epidemiological sense since the slow rise was not well detected.

The different progress of mathematically simulated epidemics are caused only by variation of one parameter: a low and a medium basic immunity I_o. Figure 6 shows the comparison of three calculated curves with three actual influenza epidemics.

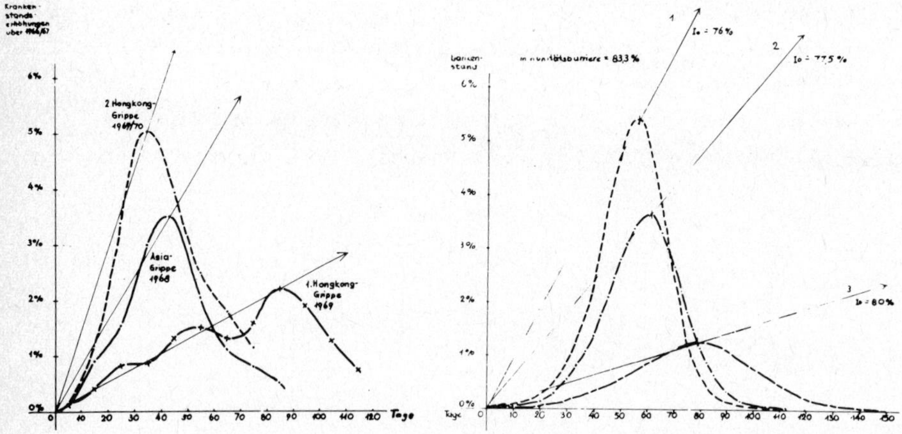

Fig. 6 Prevalences in observed (left) and simulated (right) epidemics

There are distinct similarities in the following criteria: 1. rate of rise, 2. maximal prevalence, 3. time of the maximum.

If we assume that the results of theoretical calculations may be compared to reality we can see that the immunity status of the population is important for the extent and progression of influenza epidemics.

In the model the extent and progression of the epidemic is dependent of the spreading coefficient $W_o = \frac{q}{s} \cdot N_o$, in other words it is dependent of the probable number of persons which each infectious person may infect. If the basic immunity (focal immunity) is high an epidemic may only spread slowly. If $\frac{q}{S} \cdot N$ equals unity this point is called the threshold or immunity barrier. It is identical with the "critical threshold" of KERMACK and McKENDRICK. If the spreading coefficient is smaller than unity it is impossible for an epidemic to spread since the infection chains keep getting interrupted even with massive introduction of virus. The immunity barrier of a population is not a constant. It is dependent of the possibilities of contacts or of the density and mobility of a population. If the contact number is high a high group immunity is necessary to stop the outbreak of an epidemic. If it is low very high disease rates may be expected and may be observed in reality. Examples are: kindergartens, schools, large industry and army camps (crowding groups). Thinly populated areas, on the other hand, show a definitely lower disease rate. With a low rate of contact the immunity barrier may be low. A low percentage of ill persons is sufficient to immunize the population sufficiently. For the percentage of immunes of the total population when reaching the immunity barrier

$$I_b \left[\%\right] = 100 \left(1 - \frac{1}{q}\right) \qquad (8)$$

holds.

Table 1 shows how the immunity barrier is related to the contact number:

contact number q	immunity barrier (% of population) S = 100 %	critical threshold (KERMACK and McKENDRICK) (% of susceptibles)
2	50,0 %	50,0 %
3	66,7 %	33,3 %
4	75,0 %	25,0 %
5	80,0 %	20,0 %
6	83,3 %	16,7 %
7	85,7 %	14,3 %
8	87,5 %	12,5 %
9	88,9 %	11,1 %
10	90,0 %	10,0 %

Figure 7 shows how the extent of a expected epidemic simulated by the influenza model is dependent on the difference between group immunity (focal immunity) I_o at the beginning of the epidemic and at the immunity barrier.

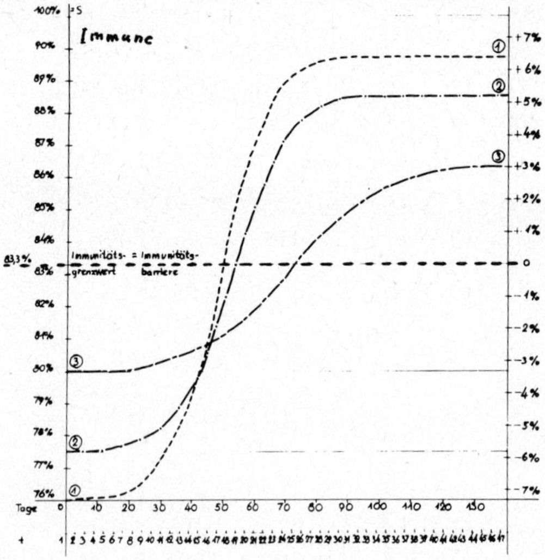

Fig. 7 The more the curve dips below the immunity barrier the higher the prevalence and the higher the proportion of immunes in the population

The simulations show that an influenza epidemic causes a collective immunization of the population against the variant of the virus causing the epidemic. After the passing of an epidemic the population as a whole is safe against a new spread of the infection by the same virus although a sizable proportion of the total population did not have the disease and therefore is not immune. Thid proportion of susceptibles (according to KERMACK and McKENDRICK) is below the "critical threshold". According to our model the proportion of immunes is "above the immunity barrier". These are two interpretations for the same fact. Practical results of the influenza model have shown that it is only useful for the simulation ofmedium and large epidemics. The limit is probably at on incidence of 6 to 7 percent. It was not possible to simulate the very small influenza epidemic of1970 to 1973 from which only few percent of the population suffered. The progress of the influenza epidemic 1974 - 1975, however, by which approximately 10 percent of adolescents and adults and approximately 10 percent of the children in the area of Niedersachsen were affected could be simulated for both groups with good agreement.
The reason for the undulating progress of influenza epidemics has not been found so far. The wavy progression of the first A/Hong Kong epidemic could be seen in figure 6 (left). Since the population did not have a specific immunity against the new virus an explosive epidemic was expected. The very slow undulating progression of the epidemic which lasted over 5 months must have had different causes. The influenza A epidemic of 1973 to 1974 which again lasted 4 to 5 months and affected only about 3 percent of the adults had also 3 maxima. At the same time a considerable infection spread of influenza B virus took place which can be seen from figure 8.

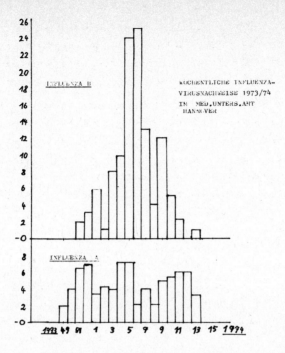

Fig. 8 Simultaneous epidemics of Influenza A and B
        - undulating progression

Medical textbooks teach us that infection with one virus excludes
the manifestation of disease by another virus: virus interference.

Literature (KRAUSS) cites examples showing that influenza B outbreaks have stopped influenza A epidemics. Using an influenza model
this phenomenon could be simulated by varying the basic immunity
I_o usually taken as constant.
By definition, it is the sum of all specific and nonspecific protection factors. The undulating process may be an interference
phenomenon of two competing viruses.
Mathematical simulation may lead to important advances in the understanding of influenza epidemics.
Mathematical simulation might also possibly explain the selection
principle of influenza A virus. In contrast to other viruses influenza A virus has a high rate of mutation chenging the behavior of
the virus towards antigens in small steps (antigen drift) and in
large steps (antigen shift).

The further a mutant shifts away from the parent family against which the population is highly immune and the larger the difference in behavior towards antigens, the higher are its chances of spread. It is unknown why the older families die off when new mutants appear.

A virus against which there is a high focal immunity has only the chance to affect a few percent of the population and attain a low proportion of infections. Another virus against whom there is a low focal immunity has the chance to gain a high infection proportion in a short time. There is a additional fact which is decisive: The genetically younger virus causes cross immunity against the older virus in the affected individuals increasing the population immunity against the older virus which completely blocks the low chances of spread. This consequent repression of an older virus can only work if it has not spread much in the population. During the maximum of the epidemic 1971/72 caused by the A/Hong Kong virus other viruses were isolated which corresponded to the succeding virus family A/England/42/72. A further spread of these viruses during the epidemic could not be observed. The behavior was different during the epidemic 1973/74. At the start of the epidemic the virus A/PtChalmers was observed against which there was as low population immunity. The simultaneously observed family A/Hannover/61/73, however, overwhelmed the genetically older family A/PtChalmers in all of Northern Europe. This repression phenomenon could possibly be simulated with a computer program which can simulate two epidemics simultaneously.

CONCLUSION

A simple model of an epidemic simulating the propagation of chain letters may be used to simulate the progress of influenza epidemics in large populations with satisfactory agreement to reality. The very short generation period of influenza A is the decisive factor for the spread of the infection and for the duration of an epidemic.

The retarded progression simulated by the model explaines the phenomenon of the slow rise of the prevalence to about one percent weeks after the first verification of the virus. The basic immunity (focal immunity) of a population is decisive for the extent and the progression of an epidemic. The concept of an immunity barrier is postulated as being dependent on the contact rates of infectious individuals of a population. The spreading coefficient is defined. It may be possible to explain complicated progressions of epidemics as interference of competing viruses and further results may be obtained by mathematically simulating the selectively better advantage of genetically younger influenza virus families.

References:

Bailey, N.T.J. : The mathematical theory of epidemics,
C. Griffin, London, Hafner, New York (1957)

Fortmann, A., Die Ausbreitung der Influenza A, Versuch der
Florian, H. Simulation durch ein mathematisches Modell,
Öff. Ges.Wes. 37, (1975), 318

Fortmann, A. Kosten-Nutzen-Analyse der Schutzimpfung
gegen Influenza,
Die gelben Hefte, XVI (1976), No. 1 (in print)

Hope Simpson, R.E. The Period of Transmission in certain epidemic
Diseases,
Lancet II, (1948), 755

Kermack, W.O., A contribution to the mathematical theory
McKendrick of Epidemics
I.Proc.Roy.Soc. A. 115, (1927), 700

Krauss, N. Die Virusgrippe: Rückblicke, Ergebnisse,
Perspektiven
Ärztl. Praxis 24, (1972), 15 - 22

Pflanz, M. Allgemeine Epidemiologie; Aufgaben, Technik,
Methoden
G. Thieme, Stuttgart (1973)

Sinnecker, H. Allgemeine Epidemiologie
G. Fischer, Jena (1971)

Wohlrab, R. Die Bedeutung der Influenza A.: Ein Beitrag
zu einer modernen Seuchenbekämpfung
Ges.Wes. u. Desinf. Heft 7, (1969)

MODEL OF RABIES CONTROL

J. Berger
Institut für Medizinische Statistik und Dokumentation
der Johannes- Gutenberg Universität Mainz, F R G

BIOLOGICAL FACTS

Rabies is an infectious disease of warm blooded animals and humans due to a neurotropic virus which is usually spread by the intake of virus contaminated saliva into a wound after the bite of a rabid animal. In the present epidemic of Western Europe which started about 1940 in the East and which shows a tendency of spreaching in a west southwesterly direction - figure 1 - (Kauker, 1975) foxes are the most important carrier of the disease. Rabies virus was isolated during 1975 in the GFR from 5718 animals of which 4180 (75 percent) were foxes and only 905 domestic animals (474 cattle, 113 dogs, 164 cats, and 154 other domestic animals).

For foxes an incubation period of 2 to 8 weeks is mentioned in the literature with a mean duration of 4 to 5 weeks. In experiments Sikes (1962) could show that sometimes in the saliva of wild living carnivores the virus is excreted before clinical symptoms appear and that the infectious period depends on the inoculation dose of the virus. The infected fox usually dies within a week after the first symptoms appear.

Jensen (1973) could demonstrate that in autumn 64 percent of the young foxes found their own living space not more than 5 km away from the environment of their parents and only 13 percent of them were recovered more than 25 km away.

Observations of Moegel et al. (1974) in the epidemic (1963 - 1967) in southern parts of Germany, of Fischer (1976) in an epidemic district of Hesse, and of Hein (1976) of the epidemic going on in southern Hesse proved that 60 percent of the newly reported cases were within a shorter distance than 5 km from a case in the preceding month and distances of new cases more than 20 km away were registered with a frequency under 5 percent.

From these facts it can be derived that the healthy as well as the infected fox does not leave his environment, that the infection is spread by biting neighbouring foxes whereby the mean conditional velocity of spread is 5 km per month - fig. 2 -, and that 80 percent

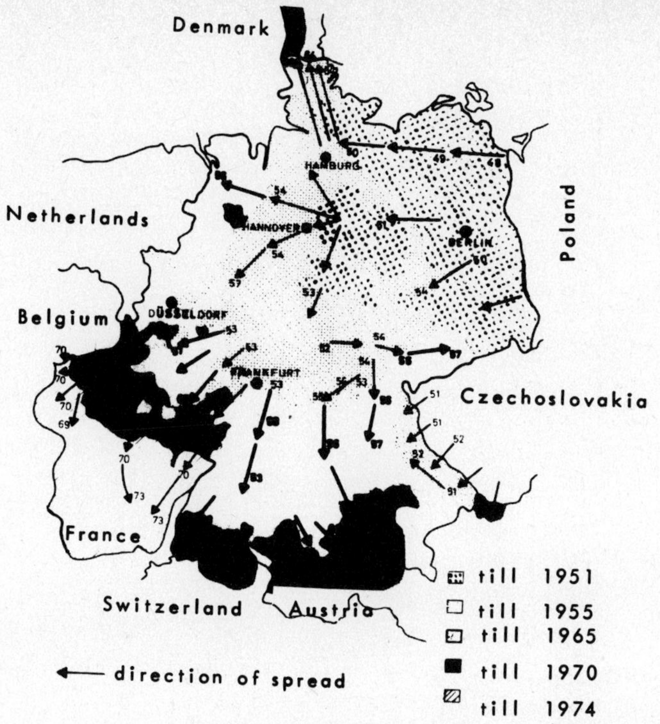

Fig.1: Spread of rabies over Western Europe
(modified according to Andral (1) and Schale (13))

of newly infected animals are observed behind the front line.

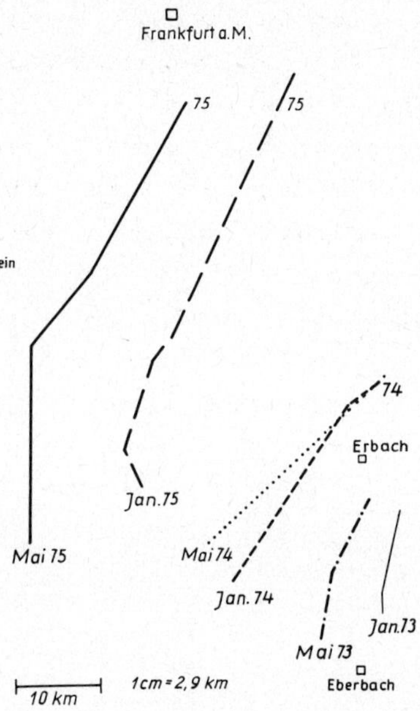

Fig. 2: Frontline of rabies epidemic in South Hesse
January 1973 - May 1975

CONTROL MEASURES AND UNANSWERED QUESTIONS

Observations in the field have shown that if the fox population density drops below the threshold of one fox per square km the spread of rabies stops.

Control measu es, however, like hunting (Kersten, 1971) or gassing operations in spring (Zinn, 1967) to lower the fox population below the critical density depend on the ecological situation and the cooperation of the local people and was therefore not successful in all regions and did not last for a long time (in some districts several years) (Bögel et al. 1974).

After the preparation by Black and Lawson (1970, 1973) of a rabies virus apathogenous to foxes but still producing immunity a new theoretical method of rabies control by oral vaccination of foxes may be possible (Regamey, 1974). A lot of experiments with marked bait have been conducted in nature since. They show an uptake rate of between 5 and 75 percent (WHO, 1975). Under conditions prevailing in the FRG Manz (1975) found an uptake rate of 60 to 70 percent independent of the fox population density.

One of the most important questions still unanswered today is: How many foxes must be immunized to stop the spread of rabies ? If such a high proportion of the population has to be immunized that the remaining susceptable individuals drop below the threshold of one fox per square km it would be necessary to immunize 80 to 90 percent of the fox population density of 3 to 5 foxes per square km. To reach such a high level of immunized foxes seems unrealistic under field conditions.

It can be assumed, on the other hand, that the presence of immune individuals in the population will slow down the spread of rabies although the number of susceptible individuals is still above the threshold. Contacts between the infectious and the immunized fox produce no new rabies case. To quantify the above theoretical considerations the following model was conceived.

MODEL

ASSUMPTION ABOUT THE POPULATION

The population is scattered over a rectangular area consisting of $6 \times 15 = 90$ equidistant grid points (figure 3). At every grid point n_{ij} ($0 \leq n_{ij} \leq 5$) foxes are located. The maximum number of foxes therefore is 450 corresponding to a population density of 5 individuals per square km. Mean population densities of 3 or 2 animals per unit square are obtained by scattering randomly 270 or 180 individuals over the area using random numbers in the interval (0,1) with the restriction that n_{ij} should not be greater than 5. Uniform distribution is used for all random numbers throughout this paper.

The program contains 5 arrays:

1. A matrix N counting the numbers of animals per grid point $n_{ij} \in N$.
2. A matrix L containing the individual latent period $l_{ijk} \in L$.
3. A matrix T containing the individual infectious period $t_{ijk} \in T$.
4. A matrix U describing the individual susceptibility $u_{ijk} \in U$.
5. A matrix A counting the individual infection age $a_{ijk} \in A$.

$i = 1,2,\ldots,6 \quad j = 1,2,\ldots,15 \quad k = 1,2,\ldots,n_{ij}$

At start every individual is assigned the infection age of -1, at the moment of infection of individual k at the grid point ij, a_{ijk} is zeroed and for every time unit elapsed its value is incremented by one.

An individual is called infectious if the condition

$$l_{ijk} \leq a_{ijk} < (t_{ijk} + l_{ijk})$$

holds.

As soon as $a_{ijk} = t_{ijk} + l_{ijk}$ the animal is considered dead and the number of individuals at grid point ij is decremented by one after reassignment of individual identify parameters in such a way that it is always the kth individual which dies.

The distribution of the latent and infectious period used listed in table 1.

The individual susceptibility is expressed by the parameter u_{ijk} defined in such way that the spread of rabies infection after a contact (bite) between an infectious (k) and a non- infected (k') animal (individual) will only take place if the random number z is less than $u_{ijk'}$. A value of u close to one hence means a high susceptibility and a value of zero an absolute immunity.

Table 1: DISTRIBUTION OF LATENT AND INFECTIOUS PERIOD OF RABIES USED IN THE MODEL

Frequency in percent

days	latent period	infectious period
3		5.0
6		43.7
9		37.7
12		11.7
15		1.3
18	0.1	0.7
21	2.0	
24	9.4	
27	23.2	
30	31.6	
33	22.5	
36	9.6	
39	1.5	
42	0.1	
mean	29.9 days	7.9 days

At the start of the simulation run a parameter ($0 \leq h \leq 1$) is chosen, determining which proportion of the population should be assigned a value of u out of the interval (.00001, .0001) and is therefore regarded as immune. The individuals defined thus a being immune are randomly scattered over the population.

SPREAD OF INFECTION

Every time unit (day) all infectious animals are counted and each of them gets an opportunity for one contact with another animal.

In order to determine the coordinates of the grid point at which this event takes place a Poisson distribution with the parameter $\lambda = 1$ is used. The probabilities for a relative movement - relative from the location of the infectious individual of 0,1,2,3,4, or 5 unit distances in the X or Y direction (fig. 3) are .3679, .3679, .1839, .0613, .0153, and .0037 respectively.

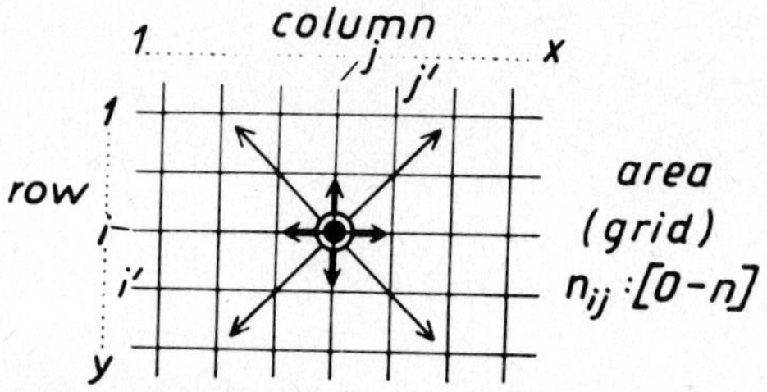

determination of the direction of migration (i', j')

$$\underline{\text{if}}\quad z_1 < e^{-\lambda} \quad \underline{\text{then}}\quad a = 0\ (b = 0)\Big\}\ l:0$$

$$\sum_{r=0}^{l-1} \frac{e^{-\lambda}\cdot\lambda^r}{r!} < z_1 < \sum_{r=0}^{l} \frac{e^{-\lambda}\cdot\lambda^r}{r!} \quad \underline{\text{then}}\quad a = l\ (b = l)\Big\}\ l:1,2,\ldots\ldots$$

$$\underline{\text{if}}\quad z_2 \leq 0{,}5 \quad \underline{\text{then}}\quad i' = i + a \quad (j' = j + b)$$

$$\underline{\text{else}}\quad i' = i - a \quad (j' = j - b)$$

Fig.3: Population structure and spread of rabies

Whereas the X and Y coordinates of the grid point of the event are chosen independently the probability for the event to happen at the next point is determined by the product of the probabilities for the X and Y coordinates. The probability for resting at the same grid point or for moving to one of the surrounding ones is $.3769^2 = .14$.

The migration movement up or down (Y- direction) or to the front or to the rear (X- direction) should happen with equal probability (.5) - no direction is preferred and is simulated by two random decisions independent of each other.

The probability whether the infectious animal will meet another one at this point or not depends on the population density (n'_{ij}) in the area and is assumed to be unity if $n'_{ij} \geq 2$ and .25 if $n'_{ij}=1$.

Using an uniformly distributed random number one of the n'_{ij} foxes is determined with which the infectious fox will have contact. Whether infection takes place is determined by the rule given above. To have comparable conditions the simulation runs are started by introducing two infectious individuals in the group 2,1 after having assigned $n_{21} = 5$.

In every run the spread of rabies is observed for 147 days the number of infected individuals is printed out for each elapsed day and the spatial distribution for each week.

RESULTS

HIGHLY SUSCEPTIBLE POPULATION

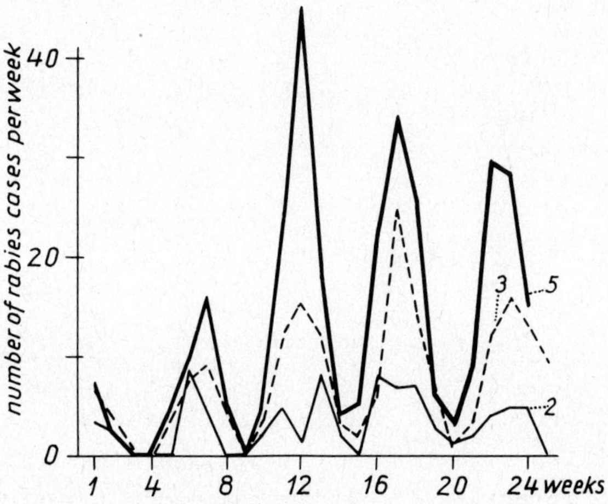

Fig.4: Weekly incidence in relation to the population density (results of the model)

Figure 4 shows the weekly incidence for a mean population density of 2, 3 and 5 nonimmunized animals per grid point. The regularly observed peaks are separated from each other by a distance corresponding to the mean latent period. This can be explained by the long latent period of rabies followed by a relatively short infectious period. In the interim period there exist none or only few infectious foxes in the population and therefore only few new cases are expected. The graph representing the monthly reported incidence, figure 5, shows a similar sawtooth pattern which, in addition, is overlapped by seasonal fluctuations due to the biology of foxes (Kauker, 1975).

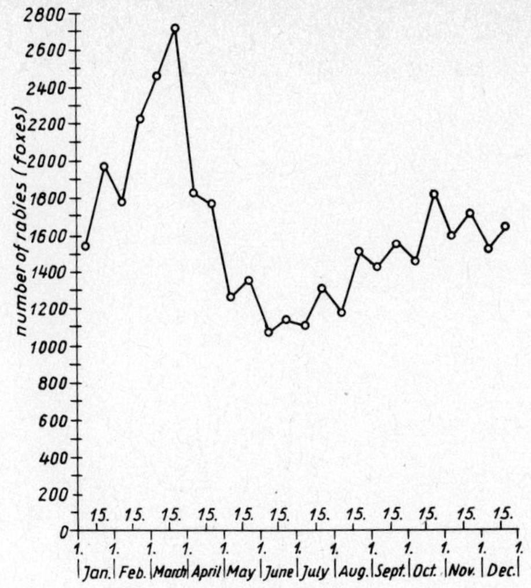

Fig.5: Monthly reported cases in the G.F.R. (13, 15)

Table 2 lists the cumulative number of infected foxes and the distance of the front line from the starting point (in grid units) at three weeks intervals for the worst cases (maximum) out of 20 epidemics simulated under the same conditions.

For the 147th day there is an additional printout of the percentage of infected foxes for the first 12 grid points (n_{11} to n_{62}), an area which can be considered as having been passed over by the rabies because no new cases had occured within it during the last 60 days.

Considering the figures of table 2 it can be demonstrated that the rabies passes over the whole area within 6 months and the whole population of foxes gets infected assuming a density of 5 foxes, per grid point. While maintaining a fully susceptible population but lowering the density the percentage of noninfected animals rises from zero to 30 percent corresponding to observations in nature after a rabies epidemic has passed over an area.

Table 2: NUMBER OF INFECTED INDIVIDUALS (I) AND SPREAD OF RABIES DEPENDENT ON THE POPULATION DENSITY AND SCHEDULE OF VACCINATION

days	mean population density per square unit	5						3				2					
	percentage of immunized foxes	0		70		80		0		60		0		45		50	
		I	II	I	II	I	II	I	II	I	II	I	II	I	II	I	II
21		12	2	7	2	4	4	13	3	7	3	7	3	6	2	6	3
42		42	5	9	4	5	4	32	6	18	6	20	4	13	3	11	5
63		61	5	11	5	5	4	47	8	21	7	25	4	13	3	14	5
84		138	9	16	5	5	4	81	8	26	7	38	5	21	6	18	5
105		214	11	22	6	5	4	121	9	29	7	54	9	29	7	25	5
126		243	11	22	6	5	4	138	10	33	7	66	9	31	7	27	7
147		319	15	26	6	5	4	175	12	33	7	83	11	36	9	33	7
minimal epidemic extent		190	9	8	5	2	1	120	9	2	1	67	8	2	1	3	1
Percentage of infected foxes in an area over which rabies has passed																	
a) relative to the total number of foxes in this area		100%		11%				94%		24%		70%		33%		23%	
b) relative to the susceptible foxes in this area		-		36%				-		58%		-		61%		46%	

I: number of infected foxes up to this time
II: frontline is ..... grid points away from the starting point

INFLUENCE OF IMMUNIZATION

If there are 5 individuals per grid point 80 percent of which are immunized corresponding to a mean population density of one susceptible fox per grid point there will be no spread of the disease under the assumptions of the model. If there are only 70 percent immunized, however, (1.5 susceptibles per grid point) the possibility of a rabies epidemic exists although the velocity of the spread is lowered. The percentage of infected foxes within the first 12 grid points is distinctly smaller both if you refer it to the total number as well as to the number of susceptible foxes present in the area at the start of the simulation. Similar relations hold for a mean population density of 2 or 3 foxes per square and a proportion of immunized foxes of 50 or 60 percent (figure 6).

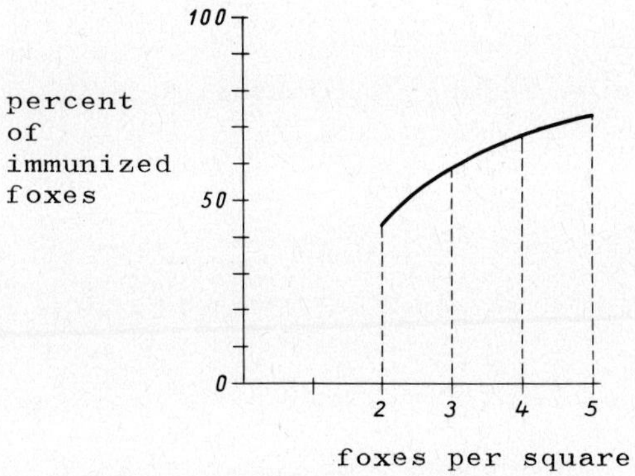

Fig.6: Required immunization schedule to prevent a rabies epidemic in relation to the population density

CONCLUSIONS

Every result of a model simulation is only as good as the assumptions made and put into the simulation as parameters. Only if the assumptions are true the results of the simulation can be valuable for practical applications. In our model the known parameters of rabies like duration of the latent and infectious period and the behaviour of foxes are used for the simulation. The weakest point of the model is the assumed contact rate between foxes. The assumption of one contact per infectious fox and per day results in a velocity of spread close to reality. The required vaccination schedule must assure that the number of remaining susceptible foxes is not very much higher than the threshold of one susceptible fox per square unit. The results show that for areas with a high fox population density vaccination only without additional control measures like hunting or gassing will not be a successful strategy to stop the spread of rabies.

REFERENCES

1 Andral, L., Toma, B.: Étude de l'epidémiologie de la rage vulpine en France (1968 - 1973)
Cah. méd. vét. 42 (1973), 203

2 Black, J. G., Lawson, F. K.: Sylvatic rabies studies in the silver fox (Vulpes vulpes)
Canad. J. Comp. Med. 34 (1970), 309

3 Black, J. G., Lawson, F. K.: Further studies of sylvatic rabies in the fox (Vulpes vulpes). Vaccination by the oral route
Canad. Vet. J. 14 (1973), 206

4 Bögel, K., Arata, A., Moegel, H., Knorpp, F.: Recovery of reduced fox population in rabies control
Zbl. Med. Vet. B. 21 (1974), 401

5 Fischer, D.: Zur Epidemiologie der Tollwut. Studien in einem endemischen Gebiet
Thesis, Mainz 1976

6 Hein, S.: Analyse der Tollwutepidemie in Süd- Hessen.
Thesis, Mainz 1976

7 Jensen, B.: Movements of the Red Fox in Denmark
Danish Rev. Yame Biol. 8 (1973)

8 Kauker, E.: Vorkommen und Verbreitung der Tollwut in Europa von 1966 bis 1974
Sitzungsberichte der Heidelberger Akademie der Wissenschaften
Springer Verlag, Heidelberg 1975

9 Kersten, W., Zinn, E.: Die Bekämpfung der sylvatischen Tollwut durch Begasung von Fuchsbauten und Abschußförderung von Füchsen
Dtsch. tierärztl. Wschr. 78 (1971), 175

10 Manz, D.: Markierungsversuche an Füchsen im Revier als Vorbereitung für eine mögliche spätere perorale Vakzination gegen Tollwut
Dtsch. Vet. Med. Gesellschaft, Bad Nauheim 12.4.1975

11 Moegel, H., Knorpp, F., Bögel, K., Arata, A., Dietz, K., Diethelm, P.: Zur Epidemiologie der Wildtollwut. Untersuchungen im südlichen Teil der Bundesrepublik Deutschland.
Zbl. Vet. Med. B, 21 (1974), 647

12 Regamey, R. H., Hennessen, W., Perkins, F. T., Triau, R.: International Symposium on Rabies (II)
S. Krager, Basel 1974

13 Schale, F. W.: Studie zur Epidemiologie der Tollwut
Thesis, Frankfurt a. M. 1975

14 Sikes, K.: Pathogenesis of Rabies in Wildlife. I. Comperative Effect of Varying doses of rabies virus inoculated into foxes and skunks
Am. J. Vet. Res. 23 (1962), 1041

15 Tierseuchenberichte des Bundesministeriums für Ernährung, Landwirtschaft und Forsten 1952 - 1975

16 W H O Report of consultions on oral vaccination of foxes against rabies
Frankfurt a. M. 1 - 3 Sept. 1975

17 Zinn, E.: Die Bekämpfung der Wildtollwut unter besonderer Berücksichtigung der Verdünnung der Fuchspopulation durch Begasung der Fuchsbaue.
Dtsch. tierärztl. Wschr. 73 (1975), 193

D i s c u s s i o n

Concerning the paper of Dietz

A. Neiss:
In your model the seasonal fluctuations are produced by the time-dependence of the parameter B. Is it not possible to get the same effects by assuming one of the other parameters as a function of time ?

K. Dietz:
Only the effect of seasonal variations of the contact rate has been investigated, because there is enough epidemiological evidence that the contact rate my undergo relatively large fluctuations. The amplitudes of the variations in the birth rate, the death rate and the removal rate are probably smaller, but it may be interesting to study also their effects. It is also conceivable to assume a seasonal variation in the vaccination rate.

Richter:
Did you analyse your basic equation system with respect to limit cycles behaviour ? At the first glance the system looks like a nonlinear oscillator.

K. Dietz:
The fact that system (1) has no limit cycle behaviour has been established previously. See Hethcote, H.W. in "Mathematical Problems in Biology: Victoria Conference", Lecture Notes in Biomathematics 2, pp 83-92.

W. Müller-Schauenburg:
Could you please sum up your experience on the coupling of your mathematical results to practical decisions within the WHO ?

K. Dietz:
My main activity in WHO is the development of mathematical models for the epidemiology and the control of major parasitic and vector-borne diseases. Some of these models are being tested in field projects specifically designed for this purpose. The experience gained contributes to the expertise required to advise member countries with respect to their disease control programmes.

Concerning the paper of Anderson

Dietz:
Have you looked into the possibility of characterising the parameter range for the stable equilibrium by an inequality which involves the reproduction rate of the parasite population for low densities ?

Anderson:
I have not as yet investigated the usefullness of defining a net reproductive rate for the parasite population since my prime concern initially has been to delineate all the parameters operating on each population in the parasite life cycle. The inequalities arising out of the stability analysis of equations 3, 4 and 5, are rather unattractive expressions and it may well be worth condensing some of the parameters into a net reproductive rate defining the number of infective stages produced by a single adult worm which achieve the adult state themselves.

Concerning the paper of Fortmann

Müller-Schauenburg:
The last example of the combined presentation of influenza A and B showed two valleys in influenza A but only one clear peak in influenza B. Do we have any chance to interpret the phenomenon in terms of an interference of only those two viruses ?

Fortmann:
During the epidemic under discussion (1973/74) the quite numerous virus assays showed almost exclusively Influenza A and B virus.

Dietz:
Is it conceivable that the multiple peaks of the influenza epidemic are due to the combination of data from geographically separated areas ?

Fortmann:
The data are from the State Medical Laboratory in Hannover covering the whole state of Lower Saxony (Niedersachsen). There is no evidence that epidemics broke out in different areas at different times showing multiple epidemic peaks. In 1969, we could observe the same

phenomenon. The prevalence curves of West-Berlin, Hannover, and Lüneburg had three peaks. Except for a few days difference the times of the three peaks coincided in all three cities.
In 1969 there was also widespread infection, usually without clinical symptoms, with influenza B virus.

Concerning the paper of Berger

Dietz:
Could you simulate a three-year cycle of rabies epidemic with your present model ?

Berger:
Until now I have not used this model to look for a cyclic behaviour of rabies but in an other model with the posibility of immunization I could demonstrate that the cycle depends on the assumption about the duration of immunity in the population.

A MARKOVIAN CONFIGURATION MODEL FOR CARCINOGENESIS

K.Schürger
P.Tautu
Institute for Documentation,
Information and Statistics
German Cancer Research Center
Heidelberg

1. INTRODUCTION

There are some biological problems in carcinogenesis which have not been satisfactorily investigated from the mathematical point of view as, for example, the role of the cell cycle, the cellular interactions, or the multistep cellular transformation process. A progress in this direction is realized in the stochastic model of T.Williams and R.Bjerknes (1971, 1972). The authors choosed the skin as cellular system because skin cancer has been extensively studied (see e.g.Bjerknes and Iversen, 1968). The cells are located on a square lattice and a one-step transformation from normal ("white") cells to malignant ("black") cells is assumed. The process starts with one black cell placed in a population of white cells; this malignant cell and all its descendants divide κ times as fast as the normal cells. Each cell divides after an exponentially distributed time displacing with equal probability one of its four neighbours. The imbedded Markov chain obtained by considering the total number of black cells at the moment a change in the configuration occurs is actually a 1-dimensional random walk with constant probabilities $p_o=\kappa/(\kappa+1)$ and $q_o=1/(\kappa+1)$, of moving right and left, respectively. The general model is, in fact, a bivariate birth-and-death process (see Bailey (1968) for a general treatment*), where the black cells are growing supercritically (see also Williams, 1974). For P.Clifford and A.Sudbury (1971) this process might be called a self-avoiding branching process, analogously to the self-avoiding random walk (Broadbent and Hammersley, 1957). A similar treatment is made by D.Downham and R.Morgan (1973 a,b) using the conjectures and results of R.Morgan and D.Welsh (1965) and J.Hammersley (1966).

In this paper we present a stochastic model for carcinogenesis taking into account the main biological phenomena mentioned above. Its initial assumptions (see A1-A5 below) are those already introduced for a "Model II" in two previous papers (Tautu, 1974; Schürger and Tautu, 1976).

---

*) This model has also been considered to have a biological prototype in the tumour growth in (Iosifescu and Tautu, 1973, p.55).

Mathematically speaking, this model might be thought as a development of the stochastic approach used by F.Spitzer (1970), R.Holley (1972), T.Liggett (1972), and T.Harris (1974) in studying physical systems of infinite particles in interaction. The method considers the transition functions as operators and uses the theory of semigroups. Motivated by A1-A5, we introduce in Section 2 an operator $\mathcal{A}$ which is proved to be the infinitesimal generator of a continuous positive semigroup $\{T_t\}_{t\geq 0}$ of contraction operators defined on some (real) Banach space of continuous real-valued functions. The properties of the linear operators and of the infinitesimal operator are presented. The existence theorem 2.2 says that the transition probabilities corresponding to the operators T_t are those of a Hunt process $\{\xi_t\}_{t\geq 0}$ which can be shown to have properties A1-A5. Section 3 is devoted to some considerations about the asymptotic behaviour of the configuration process, namely its structure and its development on the lattice. The irregularity of the configuration boundary tends to a constant value when t is large (Conjecture 3.1). Also the spatial structure tends to be "round" and "solid" for $t\to\infty$, which means that a convex geometrical figure with all sites occupied can be realized. The last conjecture suggests that $\frac{1}{t^d} S(\xi_t)$ converges to a constant for $t\to\infty$, where $S(\xi_t)$ is the number of occupied sites of a d-dimensional lattice, at time t. The computer simulations presented in Section 4 illustrate the "tumour formation" when the axioms of the configuration process $\{\xi_t\}_{t\geq 0}$ are used. The last section contains remarks about the variants and some possible developments of the considered model.

2. BASIC ASSUMPTIONS AND THE EXISTENCE THEOREM

Before formulating our basic assumptions some notation and terminology must be introduced. Let Z^d, $d\geq 1$, denote the square d-dimensional lattice, the sites (nodes) of which are the d-tuples $x=(x^1,\dots,x^d)$ of integers $x^1,\dots,x^d$. We say that a site $x=(x^1,\dots,x^d)\in Z^d$ is a neighbour of a site $y=(y^1,\dots,y^d)\in Z^d$ iff $|x^1-y^1|+\dots+|x^d-y^d|=1$. Let N_x denote the set of all neighbours of $x\in Z^d$. Clearly, $y\in N_x$ iff $x\in N_y$.

Introduce now the sets $K=\{1,\dots,k\}$, $k\geq 1$, and $S=\{1,\dots,s\}$, $s\geq 2$, kept fixed throughout. Put $X=(K\times S)\cup\{(0,0)\}$. A configuration ξ is a mapping $\xi: Z^d\to X$. The x-coordinate of ξ is given by $\xi(x)$, $x\in Z^d$. This can be interpreted as follows. Assume that an object (which will be also called B-object) is characterized by its type (or "colour") $i\in K$ and its state $j\in S$. Then a configuration ξ can be thought as a collection of objects located at the sites of Z^d : $\xi(x)=(i,j)$, $x\in Z^d$, $i\in K$, $j\in S$, means that site x is occupied by an object of type i in state j, and if $\xi(x)=(0,0)$, site x is vacant. We assume that each site of Z^d is occupied by at most one object at a time.

Put Ξ for the set of all configurations. Denote by supp(ξ), $\xi\in\Xi$, the support of ξ,

$$\text{supp}\ (\xi)=\{x\,|\,x\in Z^d, \xi(x)\neq(0,0)\},$$

and let Ξ_o denote the set of all configurations $\xi\in\Xi$ having a finite nonvoid support.

Our basic assumptions are the following (compare with Schürger and Tautu, 1976):

A1. Let a time t a B-object at $x\in Z^d$ be of colour $i\in K$ and in state $j\in S-\{s\}$. The probability that during the short time interval $(t,t+h)$ a transition from j to j+1 occurs is $a_j h+o(h)$. It is assumed that $a_j>0$, $j\in S$. This probability is supposed to be not colour-dependent.

A2. Let at time t a B-object at $x\in Z^d$ be of colour $i\in K$ and in state $s\in S$. The probability that during the short time interval $(t,t+h)$ a binary fission (division) of this B-object occurs is $d_i h+o(h)$. We assume $d_i>0$, $i\in K$.

The two new B-objects resulting from the division of their antecedent enter state $1\in S$, both taking simultaneously the same colour $i'\in K$ with probability $d_{ii'}$ such that $d_{ii'}=0$ if $i'\notin\{i-1,i,i+1\}\cap K$ and $d_{i,i-1}+d_{i,i}+d_{i,i+1}=1$, $i\in K$. (Below we put $K(i)=\{i-1,i,i+1\}\cap K$, $i\in K$.)

A3. If a B-object at $x\in Z^d$ divides, a site $y\in N_x$ is chosen with probability $\frac{1}{2d}$, all choices $y\in N_x$ being equiprobable. One of the new B-objects resulting from the binary fission is pushed to y and the B-object possibly located at y is killed and vanishes. Of course, the second B-object remains at x.

A4. No site of Z^d can be occupied by more than one B-object at a time (exclusion of multiple occupancy).

A5. The probability that at least two of the events mentioned in A1 and A2 occur in the short time interval $(t,t+h)$ is $o(h)$.

If we look at the states 1,...,s as forming a "cycle", a B-object runs along this cycle and only in state s its division and a change of type (colour) is possible. After duplication, the resulting new objects start anew to run along the cycle, beginning their life in state 1. This is an analogy with the "classical" cell cycle (see Rittgen and Tautu, 1976), where telophase, the last subphase of mitosis, is our state s. According to A2, a change of type is possible, which is interpreted as a cellular transformation (e.g.differentiation,mutation,etc.). In this sense we may say that in our carcinogenesis model a B-object of type 1 is "normal",whereas a B-object of type k is a "malignant" one,so that the type-index $i\in K$ is an indicator of the malignant transformation. According to A3,the death of a B-object is stipulated as a result of the competition process for a site. If e.g. d=2,one could also imagine that as a duplication consequence a B-object is not killed but is pushed onto a site of another 2-dimensional lattice parallel to the original Z^2.

The question arises whether it is possible to construct a Markov process with state set Ξ (i.e.configuration-valued),having properties A1-A5. Such a construction can be achieved by starting with a suitable infinitesimal generator of a semigroup of operators defined on some space. In order to make this more precise,let X have the discrete topology and give $\Xi=X^{Z^d}$ the product topology. Then Ξ is compact,metrizable, and hence has a countable base. Let $\mathcal{C}=\mathcal{C}(\Xi)$ be the family of all continuous functions $f:\Xi\to(-\infty,+\infty)$(real line). If $f\in\mathcal{C}$,the supremum norm $\|f\|$ of f is given by

(2.1) $$\|f\| = \sup_{\xi\in\Xi}|f(\xi)|.$$

With this norm, $\mathcal{C}$ is a (real) Banach space. Let $\mathcal{C}_o\subset\mathcal{C}$ denote the set of all functions only depending on finitely many coordinates. By Stone-Weierstrass theorem (see Dunford and Schwartz,1964,p.272), $\mathcal{C}_o$ is dense in $\mathcal{C}$ (i.e.for every $f\in\mathcal{C}$ and $\varepsilon>0$ there exists an $f_o\in\mathcal{C}_o$ such that $\|f-f_o\|<\varepsilon$).

For each configuration $\xi\in\Xi$ introduce configurations $\xi_y^{(j)}$,$y\in Z^d$,$j\in S$, and $\xi_{yz}^{(ii')}$,$z\in N_y$,$y\in Z^d$,$i'\in K(i)$,$i\in K$,as follows. In case $\xi(y)=(0,0)$,we put $\xi_y^{(j)}=\xi_{yz}^{(ii')}=\xi$. Now let $\xi(y)=(i_o,j_o)$,$i_o\in K$,$j_o\in S$. Then put

(2.2) $$\xi_y^{(j)}=\xi \text{ if } j\neq j_o \text{ or if } j=j_o=s$$

(2.3) $$\xi_y^{(j)}(x)=\begin{cases}(i_o,j+1), & x=y\\ \xi(x), & x\neq y\end{cases} \quad \text{if } j=j_o\neq s$$

and

(2.4) $$\xi_{yz}^{(ii')}=\xi \text{ if } i\neq i_o \text{ or if } i=i_o \text{ and } j_o\neq s$$

(2.5) $$\xi_{yz}^{(ii')}(x)=\begin{cases}(i',1), & x\in\{y,z\}\\ \xi(x), & x\notin\{y,z\}\end{cases} \quad \text{if } i=i_o \text{ and } j_o=s.$$

Thus the configurations $\xi_y^{(j)}$ represent the state transitions according to A1. We have $\xi_y^{(j)}\neq\xi$ iff in ξ the B-object located at y is in state

$j\in S-\{s\}$; in this case the change $\xi\to\xi_y^{(j)}$ represents the transition from state j to state j+1 at site y. The configurations $\xi_{yz}^{(ii')}$ are similarly interpreted.

The infinitesimal generator of a Markov process with the properties A1-A5 should now be given by

$$(2.6)\quad \mathcal{A}f(\xi) = \sum_{j=1}^{s-1} a_j \sum_{y\in Z^d}[f(\xi_y^{(j)})-f(\xi)] + $$

$$+\sum_{i=1}^{k}\sum_{i'\in K(i)} \frac{d_i d_{ii'}}{2d} \sum_{y\in Z^d}\sum_{z\in N_y}[f(\xi_{yz}^{(ii')})-f(\xi)]\ ,\ f\in\mathcal{C}_0,\xi\in\Xi.$$

It is easy to see that for fixed $j\in S, y\in Z^d, z\in N_y, i'\in K(i), i\in K$, the mappings $\xi\to\xi_y^{(j)}$ and $\xi\to\xi_{yz}^{(ii')}$, $\xi\in\Xi$, are continuous. Hence $\mathcal{A}$ is a linear operator mapping $\mathcal{C}_0$ into itself.

Let us show that the conditions of R.Holley (1972;see also Liggett, 1972) are fulfilled, which guarantee that the closure of $\mathcal{A}$ is the infinitesimal generator of a continuous positive contraction semigroup of operators $T_t:\mathcal{C}\to\mathcal{C}$, $t\geq 0$. Recall that $\{U_t\}_{t\geq 0}$ is called a <u>semigroup</u> of contraction operators on $\mathcal{C}$ (see e.g.Dunford and Schwartz,1964,p.614; Dynkin,1965,p.22) iff $U_0=I$ (the identical operator defined by $If=f$ for $f\in\mathcal{C}$);each U_t is a linear bounded operator $U_t:\mathcal{C}\to\mathcal{C}$ such that $\|U_t\|\leq 1$ (i.e. $\|U_tf\|\leq\|f\|$ for all $f\in\mathcal{C}$) and $U_{s+t}=U_sU_t$ for all $s\geq 0, t\geq 0$. $\{U_t\}_{t\geq 0}$ is called continuous iff $\lim_{t\to t_0}\|U_tf-U_{t_0}f\|=0$, for all $f\in\mathcal{C}$, $t_0\geq 0$. Also $\{U_t\}_{t\geq 0}$ is called positive if $f\geq 0$ implies $U_tf\geq 0$ for all $t\geq 0$. (Of course, $f\geq 0$ means $f(\xi)\geq 0$ for all $\xi\in\Xi$.)

The <u>infinitesimal generator</u> $\mathcal{B}$ of $\{U_t\}_{t\geq 0}$ is defined for all $f\in\mathcal{C}$ for which $\lim_{t\downarrow 0}\frac{1}{t}(U_tf-f)$ exists with respect to the norm (2.1). In this case we call the limit $\mathcal{B}f$ so that $\lim_{t\downarrow 0}\left\|\frac{1}{t}(U_tf-f)-\mathcal{B}f\right\|=0$. $\mathcal{B}$ is not defined,in general,on the whole space $\mathcal{C}$.

Define for each $y\in Z^d$ an operator Ω_y on $\mathcal{C}$ by

$$(2.7)\quad \Omega_y f(\xi) = \sum_{j=1}^{s-1} a_j[f(\xi_y^{(j)})-f(\xi)] +$$

$$+\sum_{i=1}^{k}\sum_{i'\in K(i)}\frac{d_i d_{ii'}}{2d}\sum_{z\in N_y}[f(\xi_{yz}^{(ii')})-f(\xi)],\ f\in\mathcal{C},\xi\in\Xi.$$

Obviously, Ω_y is a linear operator and

$$(2.8)\qquad \Omega_y:\mathcal{C}\to\mathcal{C},\ y\in Z^d.$$

If $f\in\mathcal{C}$ and $\xi\in\Xi$,

$$|\Omega_y f(\xi)| \le 2\|f\|\sum_{j=1}^{s-1} a_j + \sum_{i=1}^{k}\sum_{i'\in K(i)} \frac{d_i d_{ii'}}{2d}\, 2d\cdot 2\|f\| =$$

$$= 2\|f\|\Big(\sum_{j=1}^{s-1} a_j + \sum_{i=1}^{k} d_i\Big),$$

i.e. Ω_y is bounded, and

$$\|\Omega_y\| \le 2\Big(\sum_{j=1}^{s-1} a_j + \sum_{i=1}^{k} d_i\Big)\ ,\ y\in Z^d. \tag{2.9}$$

Clearly,

$$\mathcal{A}f = \sum_{y\in Z^d} \Omega_y f\ ,\ f\in\mathcal{C}_0. \tag{2.10}$$

In order to formulate other important properties of the operators Ω_y denote for each finite set $M\subset Z^d$ by $\mathcal{C}_M\subset\mathcal{C}$ the set of those functions depending only on the coordinates in M. Denote by $C_y\subset Z^d$ the cube (with center y) whose sides are of length 2 and parallel to the axes. Then

$$\Omega_y f = 0 \quad \text{if } f\in\mathcal{C}_M \text{ and } M\cap C_y=\emptyset. \tag{2.11}$$

In fact, if $f\in\mathcal{C}_M$, $f(\xi_y^{(j)})=f(\xi)$, $y\notin M$, $j\in S$, $\xi\in\Xi$. Furthermore $z\in N_y$ implies $z\in C_y$ and hence $z\notin M$. Therefore $f(\xi_{yz}^{(ii')})=f(\xi)$, $i'\in K(i)$, $i\in K$, $\xi\in\Xi$, and (2.11) is a consequence of (2.7).

Another property of Ω_y is given by

$$\Omega_y f\in\mathcal{C}_{M\cup C_y} \quad \text{if } f\in\mathcal{C}_M \text{ and } M\cap C_y\neq\emptyset. \tag{2.12}$$

This can be seen by considering configurations ξ,ξ' coinciding on $M\cup C_y$. Then clearly $f(\xi)=f(\xi')$. On the other hand, it follows from (2.2)-(2.5) that $\xi_y^{(j)}$ and $\xi'^{(j)}_y$ as well as $\xi_{yz}^{(ii')}$ and $\xi'^{(ii')}_{yz}$ coincide on $M\cup C_y$. Therefore by (2.7) $\Omega_y f(\xi)=\Omega_y f(\xi')$.

Finally let $M_1\subset M_2\subset\ldots$ be a sequence of finite subsets of Z^d such that $\bigcup_{i=1}^{\infty} M_i=Z^d$ and $C_{y_0}\subset M_1$ for some $y_0\in Z^d$. The the operator $\mathcal{A}_n$ defined on $\mathcal{C}$ by

$$\mathcal{A}_n f=\sum_{C_y\subset M_n} \Omega_y f\ ,\ f\in\mathcal{C}\ ,n=1,2,\ldots \tag{2.13}$$

is, clearly, linear, bounded (by (2.9)) and by (2.8) we have $\mathcal{A}_n:\mathcal{C}\to\mathcal{C}$. The summation in (2.13) is extended over all $y\in Z^d$ such that $C_y\subset M_n$.

Let us now show that $\mathcal{A}_n$ is the infinitesimal generator of a continuous positive semigroup $\{T_t^{(n)}\}_{t\ge0}$ of contraction operators on $\mathcal{C}$. Firstly we show

(2.14) $$\|\lambda f-\mathcal{A}_n f\| \geq \|\lambda f\| \ , \ f\in\mathcal{C} \ ,\lambda>0,n=1,2,\ldots$$

which indicates that $\mathcal{A}_n$ is dissipative (see Lumer and Phillips,1961;Liggett,1972). In fact,since Ξ is compact,there exists $\xi',\xi''\in\Xi$ such that f takes on its maximum (minimum) at $\xi'(\xi'')$. Therefore $f(\xi_y^{(j)})\leq f(\xi')$, $j\in S,y\in Z^d$,and $f(\xi_{yz}^{(ii')})\leq f(\xi'),i'\in K(i),i\in K,z\in N_y,y\in Z^d$. Therefore by (2.7), $\Omega_y f(\xi')\leq 0,y\in Z^d$. Similarly, $\Omega_y f(\xi'')\geq 0$. Consequently by (2.13), $\mathcal{A}_n f(\xi')\leq$ ≤ 0, $\mathcal{A}_n f(\xi'')\geq 0$.

Putting $g=\lambda f-\mathcal{A}_n f$ we therefore conclude $g(\xi')\geq\lambda f(\xi'),g(\xi'')\leq\lambda f(\xi'')$ and hence $\|g\|\geq\|\lambda f\|$,i.e.(2.14). It is well known that $R(\lambda,\mathcal{A}_n)=(\lambda I-\mathcal{A}_n)^{-1}$ (the inverse of $\lambda I-\mathcal{A}_n$) exists (at least) for all $\lambda>\|\mathcal{A}_n\|$ as a bounded linear operator on $\mathcal{C}$(Dunford and Schwartz,1964,p.567). $R(\lambda,\mathcal{A}_n)$ is called the resolvent of $\mathcal{A}_n$. From (2.14) we conclude

(2.15) $$\|R(\lambda,\mathcal{A}_n)\|\leq \frac{1}{\lambda} \quad \text{for all } \lambda>\|\mathcal{A}_n\| \ , \ n=1,2,\ldots$$

Furthermore

(2.16) $$R(\lambda,\mathcal{A}_n)g\geq 0 \ , \text{ if } g\geq 0,g\in\mathcal{C},\lambda>\|\mathcal{A}_n\|,n=1,2,\ldots$$

In fact,putting $f=R(\lambda,\mathcal{A}_n)g$,we have $g=(\lambda I-\mathcal{A}_n)f$. If f takes on its minimum at ξ'',$g(\xi'')\leq\lambda f(\xi'')$ as above,whence $f(\xi'')\geq 0$ implying $f\geq 0$. Now using the Hille-Yosida-Phillips theorem (see Dynkin,1965,p.30;Dunford and Schwartz,1964,pp.624,626) we deduce from (2.15) and (2.16) that $\mathcal{A}_n$ is the infinitesimal generator of a continuous positive semigroup $\{T_t^{(n)}\}_{t\geq 0}$ of contraction operators on $\mathcal{C}$. Clearly,

(2.17) $$T_t^{(n)}f = e^{t\mathcal{A}_n}f = \sum_{i=0}^{\infty}\frac{t^i}{i!}\mathcal{A}_n^i f \ , \ f\in\mathcal{C},n=1,2,\ldots$$

According to Holley (1972),the relations (2.9),(2.11),(2.12) and the properties of $\mathcal{A}_n$ imply

Theorem 2.1. There is a continuous positive semigroup $\{T_t\}_{t\geq 0}$ *of contraction operators on* $\mathcal{C}$ *such that*

(2.18) $$\sup_{0\leq s\leq t}\|T_s f-T_s^{(n)}f\|\to 0 \text{ as } n\to\infty, \ 0\leq t<\infty,f\in\mathcal{C}.$$

Put

(2.19) $$c = c(d) = [2(\sum_{j=1}^{s-1}a_j + \sum_{i=1}^{k}d_i)3^d\exp\{3^d\}]^{-1} \ , \ d\geq 1.$$

Then

(2.20) $$T_t f = \sum_{i=0}^{\infty}\frac{t^i}{i!}\mathcal{A}^i f \ , \ 0\leq t<c(d),f\in\mathcal{C}_0.$$

Furthermore $\{T_t\}_{t\geq 0}$ *is the unique contraction semigroup whose infinitesimal generator, when restricted to* $\mathcal{C}_0$, *is given by* (2.10).

Let $\mathcal{E}$ denote the family of all Borelian subsets of Ξ. It has been remarked that Ξ is a compact space having a countable base. Let $P_t(\xi,A)$, $\xi\in\Xi, A\in\mathcal{E}, t\geq 0$, denote the transition probabilities corresponding to $\{T_t\}_{t\geq 0}$, according to the Riesz representation theorem (see Dunford and Schwartz, 1964, p.265). Observe that Ξ is a compact Hausdorff space and consider for fixed $t\geq 0$ and $\xi\in\Xi$ the functional $f\to T_t f(\xi), f\in\mathcal{C}$. Then we have (see Blumenthal and Getoor, 1968, p.46)

Theorem 2.2. (Existence theorem). *There exists a Hunt process* $\{\xi_t\}_{t\geq 0}$ *with state space* $(\Xi,\mathcal{E})$ *and transition probabilities* $P_t(\xi,A)$, *i.e. among others,* $\{\xi_t\}_{t\geq 0}$ *is a time-homogeneous quasi-left-continuous (on* $[0,\infty)$) *strong Markov process whose all paths are right-continuous and have left-hand limits.*

We write $P_\xi\{\xi_t\in A\}$ instead of $P_t(\xi,A), \xi\in\Xi, A\in\mathcal{E}, t\geq 0$.

One can now prove that our configuration-valued process $\{\xi_t\}_{t\geq 0}$ has the properties A1-A5. The process $\{\xi_t\}_{t\geq 0}$ has the property of "preservation of ergodicity". Define the shift σ_a $(a\in Z^d)$ by

$$(2.21)\qquad \sigma_a(\xi)(x)=\xi(x-a)\ ,\ \xi\in\Xi, x\in Z^d, a\in Z^d.$$

Furthermore we put for each mapping $f:\Xi\to(-\infty,+\infty)$ being measurable with respect to $\mathcal{E}$,

$$(2.22)\qquad \sigma_a(f) = f\circ\sigma_{-a}\ ,\ a\in Z^d.$$

Observe that $\sigma_a A\in\mathcal{E}$ if $A\in\mathcal{E}, a\in Z^d$.

To apply a result of R.Holley (1972) we need two additional properties of the operator $\Omega_y, y\in Z^d$. We have first

$$(2.23)\qquad \Omega_{y+a}\sigma_a f = \sigma_a\Omega_y f\ ,\ a,y\in Z^d, f\in\mathcal{C}.$$

In order to prove it, observe that

$$(2.24)\qquad \sigma_{-a}(\xi^{(j)}_{y+a}) = (\sigma_{-a}\xi)^{(j)}_y\ ,\ y,a\in Z^d, j\in S, \xi\in\Xi,$$

and

$$(2.25)\qquad \sigma_{-a}(\xi^{(ii')}_{y+a,z+a}) = (\sigma_{-a}\xi)^{(ii')}_{yz}\ ,\ z\in N_y, y,a\in Z^d, i'\in K(i), i\in K, \xi\in\Xi.$$

Application of (2.7) together with (2.22),(2.24),and (2.25) yields (2.23).

Furthermore we have

$$(2.26)\qquad \Omega_y(fg) = g\Omega_y f\ ,\ y\in Z^d, f\in\mathcal{C}, g\in\mathcal{C}_M, M\cap C_y=\emptyset.$$

In (2.26) fg is defined as $(fg)(\xi)=f(\xi)g(\xi),\xi\in\Xi$. Remark that $z\in N_y$ implies $z\in C_y$. If $\xi\in\Xi$, by (2.2)-(2.5) $\xi_y^{(j)}$ and $\xi_{yz}^{(ii')}$ coincide with ξ outside of C_y and since $g\in\mathcal{C}_M$, $M\cap C_y=\phi$, we have $g(\xi_y^{(j)})=g(\xi_{yz}^{(ii')})=g(\xi)$, $j\in S$, $i'\in K(i)$, $i\in K$, $z\in N_y$. Now the application of (2.7) immediately yields (2.26).

Call a probability measure μ on $\mathcal{E}$ <u>shift</u> <u>invariant</u> iff

$$\mu(\sigma_a A) = \mu(A) \ , \ a\in Z^d, A\in\mathcal{E}. \tag{2.27}$$

It is easy to see that μ is shift invariant iff

$$\int_\Xi \sigma_a(f)(\xi)\mu(d\xi) = \int_\Xi f(\xi)\mu(d\xi) \ , \ a\in Z^d, f\in\mathcal{C}. \tag{2.28}$$

Clearly, (2.28) is necessary for μ to be shift invariant. To prove that (2.28) is sufficient, first apply a construction of Dynkin (1965, <u>1</u>, p.138) and then Lemma 0.3 (Dynkin, 1965, <u>2</u>, p.203; see also Blumenthal and Getoor, 1968, p.6 and Meyer, 1966, p.11). Let $\mathcal{M}$ denote the set of all shift invariant probabilities. A probability $\mu\in\mathcal{M}$ is called <u>ergodic</u> for the shifts σ_a if $\sigma_a(A)=A$, for all $a\in Z^d$, implies that $\mu(A)$ is either zero or one.

If μ is a probability measure on $\mathcal{E}$, define the measure $T_t^*\mu$ by

$$T_t^*\mu(A) = \int_\Xi P_\xi\{\xi_t\in A\}\mu(d\xi) \ , \ A\in\mathcal{E}, t\geq 0. \tag{2.29}$$

$T_t^*\mu(A)$ is the probability that the configuration-valued process is in A at time t provided that it starts at a random initial configuration, being distributed according to μ. It is easy to see (compare with the remarks after (2.28)) that (2.29) is equivalent to

$$\int_\Xi f(\xi)T_t^*\mu(d\xi) = \int_\Xi T_t f(\xi)\mu(d\xi) \ , \ f\in\mathcal{C}, t\geq 0, \tag{2.30}$$

where $\{T_t\}_{t\geq 0}$ is the semigroup in Theorem 2.1. Since $T_t 1=1$ by (2.20), $T_t^*\mu$ is again a probability measure for $t\geq 0$. Now it follows (Holley, 1972)

<u>*Theorem 2.3.*</u> (Preservation of ergodicity) *Let* $\mu\in\mathcal{M}$ *be ergodic for the shifts* σ_a. *Then* $T_t^*\mu$ *is ergodic for all* $t\geq 0$.

3. <u>ASYMPTOTIC BEHAVIOUR</u>

In this section we consider the asymptotic behaviour of the process $\{\xi_t\}_{t\geq 0}$. The axiom A3 suggests that the configuration process should have the following property : Consider sites $x_1, x_2\in Z^d$ which are not occupied at time t_1 such that at this epoch x_1 has more occupied neighbours than x_2. Then the probability that x_1 will be occupied during a fixed time interval $(t_1,t_2], t_1<t_2$, should be greater ("in the average") than the corresponding probability for x_2. This should have, in turn, a smoothing effect on the shape of the set of occupied sites. In order to make this idea more precise we introduce some additional notation and terminology. Assign to each $x=(x^1,\ldots,x^d)\in Z^d$ a cube Q_x defined by

$$(3.1) \qquad Q_x=\{y=(y^1,\ldots,y^d)\in R^d \mid x^i-\frac{1}{2}<y^i\le x^i+\frac{1}{2},\ 1\le i\le d\},$$

where R^d is the d-dimensional Euclidean space. Call $z\in R^d$ occupied if the center x of the (unique) cube Q_x to which z belongs is occupied. Let $B(\xi)$, $\xi\in\Xi$, denote the boundary (in the usual topological sense) of the union of all $z\in R^d$ occupied in ξ, and in case d=2 denote by $\ell[B(\xi)]$ its length. Let $S(\xi)$ denote the number of sites in supp(ξ). According to the above argumentation, $B(\xi_t)$ should have a self-stabilizing property preventing $B(\xi_t)$ to become too "irregular". A natural measure of the irregularity of $B(\xi)$, $\xi\in\Xi$, is the <u>crinkliness</u> proposed by D.Mollison (1972). The idea behind consists in comparing $\ell[B(\xi)]$ with the length of the boundary of a square array having the same area as ξ. More precisely, the crinkliness $C(\xi)$ of $\xi\in\Xi_o$ is defined, in case of d=2, as

$$(3.2) \qquad C(\xi)=\frac{\ell[B(\xi)]}{4\sqrt{S(\xi)}},\quad \xi\in\Xi_o.$$

This implies that $C(\xi)\ge 1$ for all $\xi\in\Xi_o$, and $C(\xi)=1$ iff the set of all $z\in R^d$ occupied in ξ is a square. Now we formulate

<u>*Conjecture 3.1.*</u> *Assume* d=2. *Let the process* $\{\xi_t\}_{t\ge 0}$ *start in* $\xi\in\Xi_o$. *Then there exists a positive constant* α *(depending on* a_j, d_i *and* d_{ii}*, but possibly not on* ξ*) such that*

$$(3.3) \qquad \lim_{t\to\infty} C(\xi_t)=\alpha, \textit{almost everywhere } (P_\xi).$$

This conjecture -which was confirmed by the computer simulations- is possibly very difficult to be proved. One can prove that starting with $\xi\in\Xi_o$ one has $P_\xi\{\xi_t\in\Xi_o \text{ for all } t\ge 0\}=1$.

The second conjecture concerns $S(\xi_t)$ in general case $d\ge 1$. The above considerations suggest that ξ_t should tend to be "round" and "solid" for large t. The results of our computer simulations (see the figures in the next section) clearly show this tendency. D.Richardson (1973) studied in his paper a class of stochastic spatial processes with a similar behaviour.

Let $\tau_1(x)$, $x\in R^d$, denote the first epoch when x is occupied. Consider now (Richardson, 1973) the random variable $\tau_h(x)=h\tau_1(\frac{x}{h})$, $h>0$. (If $h>0$

is small, $\tau_1(\frac{x}{h})$ will be large, for $x\neq 0$, but the factor h compensates this tendency.) If E_ξ denotes the formation of expectation with respect to $P_\xi(\cdot)$, it follows from Richardson's method that under certain conditions

$$(3.4) \qquad \lim_{h\downarrow 0} E_\xi[\tau_h(x)]=N(x) \quad , \ \xi\in\Xi_o$$

exists for all $x\in R^d$ and defines a norm on R^d, equivalent to the Euclidean norm $\|x\|$ of $x\in R^d$. This means that there exist constants $0<a_1\leq a_2$ such that

$$(3.5) \qquad a_1\|x\|\leq N(x) \leq a_2\|x\| \quad , \ x\in R^d,$$

which implies that also N(x) generates the Euclidean topology (see e.g. Dieudonné, 1969, p.106). The main theorem in Richardson's paper (p.521) is now that under certain conditions, for all $\varepsilon>0$,

$$(3.6) \quad \{x|x\in R^d, N(x)\leq(1-\varepsilon)t\}\subset\{x|\tau_1(x)\leq t\}\subset\{x|x\in R^d, N(x)\leq(1+\varepsilon)t\}$$

with a probability $\pi(\varepsilon,t)$ such that

$$\lim_{t\to\infty} \pi(\varepsilon,t)=1 \quad , \ \varepsilon>0.$$

The event in (3.6) means that at time t all points $x\in R^d$ of the closed ball with radius $(1-\varepsilon)t$ are occupied and all occupied points are contained in the closed ball of radius $(1+\varepsilon)t$. The word "radius" refers to norm $N(\cdot)$.

This is the precise mathematical statement of "round" and "solid". Now we formulate

Conjecture 3.2. *For all $\xi\in\Xi_o$ and $x\in R^d$ the limit in (3.4) exists and defines a norm equivalent to the Euclidean norm. Furthermore for all $\varepsilon>0$,*

$$(3.7) \quad \lim_{t\to\infty} P_\xi[\{x|N(x)\leq(1-\varepsilon)t\}\subset\{x|\tau_1(x)\leq t\}\subset\{x|N(x)\leq(1+\varepsilon)t\}]=1 \quad , \ \xi\in\Xi_o.$$

The proof of this conjecture will be presented in a forthcoming paper. Conjecture 3.2 leads to

Conjecture 3.3. *For all $\xi\in\Xi_o$ there exists a constant $\beta>0$ such that*

$$(3.8) \qquad \lim_{t\to\infty} \frac{S(\xi_t)}{t^d} = \beta, \textit{almost everywhere } (P_\xi).$$

This is suggested by the fact that if n(t) denotes the number of sites of Z^d which are contained in $\{x|N(x)\leq t\}$, there exists a constant $\beta>0$ such that

$$(3.9) \qquad \lim_{t\to\infty} \frac{n(t)}{t^d} = \beta.$$

4. COMPUTER SIMULATIONS

In order to simulate our model on a computer, a program in APL has been written (see e.g. Katzan, 1970). The choice of APL is motivated by its ideal suitability for mathematical purposes. For example, operations on matrices can be easily programmed. The program was run on IBM 370/158 machine of the German Cancer Research Center, Heidelberg.

The calculation of the sequence of random numbers $y_o, y_1, y_2, \ldots$ is based on the recursion

$$(4.1) \qquad y_{i+1} \equiv a y_i \pmod{2^m}\ , \ i=0,1,2,\ldots$$

where $y_o \equiv 1 \pmod 4$ and $4a = 2^m \gamma$, where $\gamma = \frac{1}{2}(\sqrt{5} - 1)$ is the Golden section number (see Ahrens et al., 1970; Ahrens and Dieter, 1972). We choose m=27 and $a = 2^{25}\gamma \approx 20737781$, where the latter integer is $\equiv 5 \pmod 8$ and as close as possible to $2^{25}\gamma$. It is easy to see that $2^{27} \times 20737781$ has not more than 16 decimals. From $y_o, y_1, y_2, \ldots$ one easily derives random numbers which are either uniformly distributed in (0,1) or exponentially distributed with a given parameter.

In the following five pictures we present one of our simulation experiments where we choosed $S=\{1,2\}$ and $K=\{A,\ldots,E\}$. In these figures C2, for example, will denote a B-object of type C in state 2. We started with a single cell of type A in state 1 and we choosed the parameters $a_1=1$ and $(d_1,\ldots,d_5)=(7,8.4,9.8,9.8,11.2)$, together with the transition probability matrix for the changes of types,

$$\underline{P} = \begin{pmatrix} 0.7 & 0.3 & 0 & 0 & 0 \\ 0.1 & 0.6 & 0.3 & 0 & 0 \\ 0 & 0.1 & 0.6 & 0.3 & 0 \\ 0 & 0 & 0.1 & 0.5 & 0.4 \\ 0 & 0 & 0 & 0.3 & 0.7 \end{pmatrix}.$$

Fig.1 shows the configuration obtained after 105 changes (that is, changes of state and divisions) and Fig.2 shows the growth of the cell population after another 400 changes.

Let us look at Fig.2 in order to explain the numerical data produced by the simulation. The numbers in the first line, namely 505, 17 , and 14.20004041, give the number of changes, the number of computer prints and, respectively, the time t at which the 505-th change occurs. Below the produced configuration, number 57 represents the total number of B-objects in this configuration, whereas 42 is the total length of the boundary (calculated by a recursive procedure). The vector (2,1,9,17,28) gives the numbers of B-objects of different types (say,two "normal" cells and 28 "malignant" cells) and the vector (53,4) gives the numbers of B-objects in different states. The corresponding frequencies of types and states are given by the next two vectors. If we assume that state 2 represents the mitotic phase,the mitotic index is given by the last number of the second vector : it equals 7.017% in this configuration.

```
105  11  8.826760725

   D1 D1
C1 D1 D1
C1 E1 E1 E1 C1
   C1 C1 D1 C1 E1 E1
   D1 C1 D1
   C1 C1 C1

22  26
0  0  10  7  5
22  0
0  0  0.4545454545  0.3181818182  0.2272727273
1  0
3.541441132  1.385804656
```

Fig.1. A configuration after 105 changes (see the text)

```
505  17  14.20004041

      D1 D1 E1
      E1 E1 D1 D1
C1 D1 D1 D2 D1 D1 D1
D2 D1 D1 C2 E1 D1 E1 E1
A1 C1 C1 C1 E1 E1 D1
A1 B1 C1 C1 E1 E1 E1 C1
E1 E1 E1 E1 E1 E1    C1
D2 E1 E1 E1 E1 E1
E1 E1 E1 E1 D1
            E1
            E1

57  42
2  1  9  17  28
53  4
0.0350877193  0.01754385965  0.1578947368  0.298245614
      0.4912280702
0.9298245614  0.0701754386
3.537563995  1.390758975
```

Fig.2. A configuration after 505 changes
(the same computer experiment as above)

In the last line of Fig.2,the number 3.537563995 is equal to t^2 divided by the total number of B-objects (see Conjecture 3.3) and is called the growth quotient. The second number gives the crinkliness (formula (3.2)) which,in this case,equals $\frac{42}{4\sqrt{57}}$.

The next three pictures show the configurations with 91,167 and, respectively,236 cells at t=18.42,22.86 and,respectively,25.57. The numbers of the B-objects of type D and E ("precancerous" and "malignant" cells) are continuously increasing and -because the reversible transformations are assumed- in the population also exist B-objects of type A,B, and C.

```
1005  22  18.42440298

   D1    E1 E2 E1 E1 D1 E1
   D2 E2 E1 D1 E1 E1 D1 E1 D1
      E1 E2 D1 E1 E1 E1 E1 E1
      E1 D1 D1 E1 D2 E1 D1 D1 E2
   B1 D2 D1 D1 D1 D1 D1 D1 D1 B1 B1
E2 E1 D1 E1 E1 E1 E1 D1 D1 D1 C1
      E1 E2 D1 E1 E1 C1 E1 D1
      D1 D1 D1 E1 E1 E1 E1 D1
E1 E1 E1 D1 E2 D1 D1 E1 D1
   D1 D1 E1 D1 E2 E1 E1
   D1 D1       E1 E1

91  50
0  3  2  39  47
80  11
0  0.03296703297  0.02197802198  0.4285714286  0.5164835165
0.8791208791  0.1208791209
3.730314564  1.467598771
```

Fig.3. The produced configuration after 1005 changes

```
2005  32  22.86391257

               E1 E1 E2
               D1 D1 D1 E1    E1
   D1    D2 E1 D1 D1 D1 D1 E1 D1 E1 E1
E1 E1 E1 D1 E1 E1 E1 E1 E1 E1 D1 E1 C1 C1
         E1 E1 E1 E1 D1 D1 D1 E1 B1 E1 E1 E1
         E1 E1 E2 E1 E1 E1 D1 E2 B2 D1 D1 D2
         E1 D1 D1 E1 E1 E1 D1 C1 E1 E1 D1    A2
      E1 E1 E1 E1 E1 E1 E1 D1 E1 E1 D1 D1 D1 A1 A1
      E2 E1 E1 E1 D1 D2 E1 D1 E1 E1 D2 D1 D1 E1
      D1 D1 E1 C1 C1 C1 C1 D1 E1 E1 E1 D1 E1 E1
         C1 C1 D1 D1 C1 E2 D1 E1 E1 E1 D1 E1
   E1 E1 E1 D1 C1 C2 D1 E1 E1 E1 D1 D1 D1 E1
   E1    D1 D1 D1 C1 C1 C1 D1 D1 E2 E1 E2
            E1 D1 C1 C2 D1 E1 D1 D1 E1 E1 E1
            E1 E1
      E1 E1 D1 E1

167  88
3  2  17  56  89
152  15
0.01796407186  0.0119760479  0.1017964072  0.3353293413
      0.5329341317
0.9101796407  0.08982035928
3.130290408  1.702411112
```

Fig.4. The produced configuration at step 2005

```
3005  42  25.57352056

                  D1
                  D1
                  D1 D1 D1 E1
                  E1 D1 E1 E1
            D1 D1 E1 E1 E1 D1 E1 E1 E1
         E1 E1 E1 E1 E1 E2 E1 E1 D1 D1 E1 E1 E1
D1 C1 D1 E1 E1 E1 E1 E1 E1 E1 E1 E1 D1 E1 E1 E1 E1
D1 C1 D1 E1 E1 E1 D2 D2 D1 D1 E1 D1 B1 B1 D1 D1 D1 E1
         E1 E1 E1 E1 E1 E2 E1 E1 D1 C1 C1 C1 E1 E2 D1
            E1 E1 E1 E1 E1 D1 D1 D1 C1 C1 C2 E1 A1 A1 B1 B1 B1
         E1 E2 E1 E1 E1 E1 E1 E2 E1 E1 D1 C1 E2 C1 E1 A1 B1 B1
         D1 D1 E1 E1 E1 D1 D2 D1 E1 E1 D1 C1 C1 C1 D1 A1
   C1 C1 D1 D1 D2 D1 E1 E1 E1 E1 D1 E1 E1 E1 D1 D1 D1
      E1 E1 D1 D2 E1 E1 C1 C1 E1 D1 D2 C1 E1 E1 D2 E1 E1
   D1 D1 D1 E1 E1 E1 E1 D1 D1 E1 E1 C2 C1 D1 D1 E1
      E2 E1    E1 E1 B1 D1 D1 E1 E1 E1 D1 D1 D2 D1 E1
      E1 E1 E1 E1 D1 B1 D1 C1 C1 B1 E1 D1 D1 D2 D1 E1
         E1 E1    D1 E1 A1 A1 B1    B1 B1 D1 D1    E1
            E1 D1 E1 E1                D1
            E1 D1 E1
            E1    E1

236  112
6  13  22  75  120
218  16
0.02542372881  0.05508474576  0.09322033898  0.3177966102
     0.5084745763
0.9237288136  0.07627118644
2.771207431  1.822644754
```

Fig.5. The produced configuration at step 3005

The growth coefficient decreases slowly (from 3.73 to 2.77) but the irregularities of the structure are evident.

5. CONCLUDING REMARKS

Our axiom A3 clearly implies that in the process $\{\xi_t\}_{t\geq 0}$ starting at some $\xi\in\Xi_o$ the number of B-objects is increasing. It is easy to modify A3 in a natural way such that the resulting process has a constant number of objects. To this end, fix a nonvoid finite set $Z_o\subset Z^d$ such that $N_x\cap Z_o\neq\emptyset$ for all $x\in Z_o$. Let the axiom A3′ differ from A3 only by the assumption that if a B-object at some $x\in Z_o$ divides, some $y\in N_x\cap Z_o$ is chosen such that all choices are equiprobable. The modified process is assumed to start at some $\xi\in\Xi_o$ such that $\mathrm{supp}(\xi)=Z_o$. This has been called Model I in (Schürger and Tautu, 1976).

The axiom A1 can be weakened by allowing a_j to be also colour-dependent, i.e. the probability that a B-object of colour $i\in K$ in state $j\in S-\{s\}$ divides during $(t,t+h)$ is $a_{ij}h+o(h)$. Conjecture 3.2 should cover this case, too, provided $a_{ij}>0, i\in K, j\in S$.

An interesting variant of this model originates in the observation that in A1 and A2 both a_j and d_i depend neither on $\xi\in\Xi$ nor on $y\in Z^d$. For its biological consequences, the following situation is to be considered. Let $\phi_i,\psi_j:[0,\infty)\to(0,\infty)$ be bounded, and such that

$$\lim_{u\to\infty}\phi_i(u)=0\ ,\ i\in K,$$

$$\lim_{u\to\infty}\psi_j(u)=0\ ,\ j\in S-\{s\}.$$

Assume

(5.1) $$d_i = d_i(y) = \phi_i(\|y\|) \ , \ y \in Z^d, i \in K,$$

(5.2) $$a_j = a_j(y) = \psi_j(\|y\|) \ , \ y \in Z^d, j \in S-\{s\},$$

where $\|y\|$ denotes the Euclidean norm of y. (For the sake of simplicity we assume that d_i and a_j are not configuration-dependent.) For example, (5.1) means that the division rate of a B-object now depends on its distance from the origin. It is easy to see that all essential properties of $\mathcal{A}$ and Ω_y still hold - but the bound in (2.9) changes, of course. Hence a Hunt process with the desired properties still exists and also Theorem 2.2 and Theorem 2.8 given by T.Liggett (1972) are applicable.

Our Conjecture 3.3 is, obviously, not valid in the present case. In fact, it is even probable that if the functions ϕ_i and ψ_j are small enough for large arguments, the number of B-objects of the process $\{\xi_t\}_{t\geq 0}$ starting at $\xi_o \in \Xi_o$ should remain bounded with probability 1.

REFERENCES

1. Ahrens,J.H.,Dieter,U.(1972).Computer methods for sampling from the exponential and normal distributions. Comm.ACM 15,873-882
2. Ahrens,J.H.,Dieter,U.,Grube,A.(1970).Pseudo-random numbers.A new proposal for the choice of multipliers. Computing 6,121-138
3. Bailey,N.T.J.(1968).Stochastic birth,death and migration processes for spatially distributed populations. Biometrika 55,189-198
4. Broadbent,S.R.,Hammersley,J.M.(1957).Percolation processes.I.Proc. Cambridge Philos.Soc.53,629-641
5. Blumenthal,R.M.,Getoor,R.K.(1968).Markov Processes and Potential Theory. New York:Academic Press
6. Clifford,P.,Sudbury,A.(1971).Some mathematical aspects of two-dimensional cancer growth. Unpubl.paper
7. Dieudonné,J.(1969).Foundations of Modern Analysis. New York:Academic Press
8. Downham,D.Y.,Morgan,R.K.B.(1973).Growth of abnormal cells. Nature 242,528-530
9. Downham,D.Y.,Morgan,R.K.B.(1973 b).A stochastic model for a two-dimensional growth on a square lattice. Bull.Intern.Statist.Inst.45(Book1), 324-331
10. Dunford,N.,Schwartz,J.T.(1964).Linear Operators.Part I. New York: Intersci.Publ.
11. Dynkin,E.B.(1965).Markov Processes.Vol.1,2. Berlin-Göttingen-Heidelberg:Springer-Verlag
12. Hammersley,J.M.(1966).First-passage percolation. J.Roy.Statist.Soc. Ser.B 28,491-496
13. Harris,T.E.(1974).Contact interactions on a lattice. Ann.Probability 2,969-988

14.Holley,R.(1972).Markovian interaction processes with finite range interactions. Ann.Math.Statist.43,1961-1967

15.Iosifescu,M.,Tautu,P.(1973).Stochastic Processes and Applications in Biology and Medicine. Berlin-Heidelberg-New York:Springer-Verlag

16.Katzan,H.(1970).APL Programming and Computer Techniques. New York:Van Nostrand-Reinhold

17.Liggett,T.M.(1972).Existence theorems for infinite particle systems. Trans.Amer.Math.Soc.115,471-481

18.Lumer,G.,Phillips,R.S.(1961).Dissipative operators in a Banach space. Pacific J.Math.11,679-698

19.Meyer,P.A.(1966).Probability and Potentials. Waltham,Mass.:Blaisdell

20.Mollison,D.(1972).Conjecture on the spread of infection in two dimensions disproved. Nature 240,467-468

21.Morgan,R.W.,Welsh,D.J.A.(1965).A two-dimensional Poisson growth process. J.Roy.Statist.Soc.Ser.B 27,497-504

22.Richardson,D.(1973).Random growth in a tessellation. Proc.Cambridge Philos.Soc.74,515-528

23.Rittgen,W.,Tautu,P.(1976).Branching models for the cell cycle. This volume

24.Schürger,K.,Tautu,P.(1976).Markov configuration processes on a lattice. Rev.Roumaine Math.Pures Appl.21,233-244

25.Spitzer,F.(1970).Interaction of Markov processes. Advances in Math. 5,246-290

26.Tautu,P.(1974).Random systems of locally interacting cells (abstr.) Advances in Appl.Probability 6,237

27.Williams,T.(1971).Working paper. Unpubl.work

28.Williams,T.(1974).Evidence for super-critical tumour growth (abstr.) Advances in Appl.Probability 6,237-238

29.Williams,T.,Bjerknes,R.(1971).A stochastic model for the spread of an abnormal clone through the basal layer of the epithelium. Unpubl. paper

30.Williams,T.,Bjerknes,R.(1972).Stochastic model for abnormal clone spread through epithelial basal layer. Nature 236,19-21

BRANCHING MODELS FOR THE CELL CYCLE

W.Rittgen

P.Tautu

Institute for Documentation,
Information and Statistics
German Cancer Research Center
Heidelberg

1. INTRODUCTION

The definition of the cell cycle as the continuous and dependent sequence of events which mark the life of a cell from its birth to its division into two daughter cells, clearly shows the extreme importance of this repetitive process for fundamental biological studies. Essentially, the cell cycle mimics in miniature many of the processes that govern the development of an organism or the differentiation process at a structural level (Mitchison, 1973). In a certain sense the cancer problem is the cell cycle problem. The action of many pathogenic agents as well as the action of the agents with therapeutic value is exerted at certain points of this cycle.

Long time ago this process has been rudimentarily known: the separation process called mitosis (with its four subphases: prophase, metaphase, anaphase, and telophase) was the single identifiable state of the cell life, preceded by a long (about 95% of the whole lifespan) silent interphase (Fig.1).

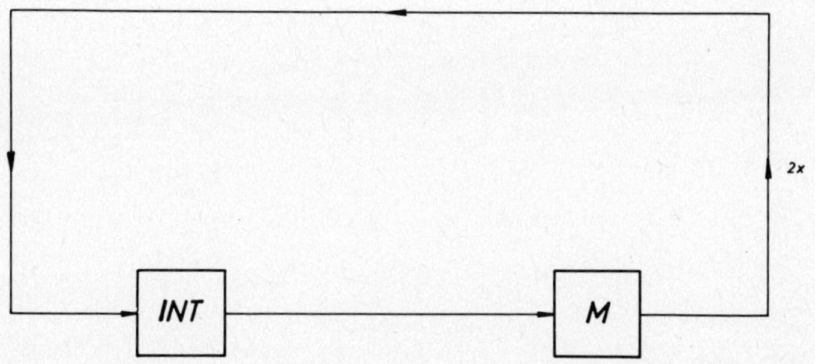

Fig.1. The cell lifespan with two phases
(Int.: interphase, M: mitosis)

The idea of many other stages originates in the experiments studying the generation time - e.g. the lifespan of bacteria (Kelly and Rahn,

1932;Rahn,1932) - and its distribution. D.G.Kendall (1948) considered first the cellular growth process as a sum of $k \geq 2$ stochastic subprocesses (called intrinsic processes in Iosifescu and Tautu,1973,p.60). These subprocesses should be simultaneous or sequential (Fig.2) and then the distribution of the generation time will be the same as the distribution of the finite sum of iid random variables,that is,

$$S_k = \frac{u_1}{1} + \frac{u_2}{2} + \ldots + \frac{u_k}{k}$$

in case of simultaneous phases (interpreted as the replication of cell components,perhaps genes),and

$$S_k = \frac{u_1}{b_1} + \frac{u_2}{b_2} + \ldots + \frac{u_k}{b_k}$$

in case of successive phases,where the length of phase $i, 1 \leq i \leq k$,has a negative exponential probability density $b_i e^{-b_i t}$ (see Kendall,1952). Consequently,the interphase duration is distributed as the sum of k-1 independent phases,each of them distributed like the mitotic (k-th) phase (see also Harris,1963,p.114,156).

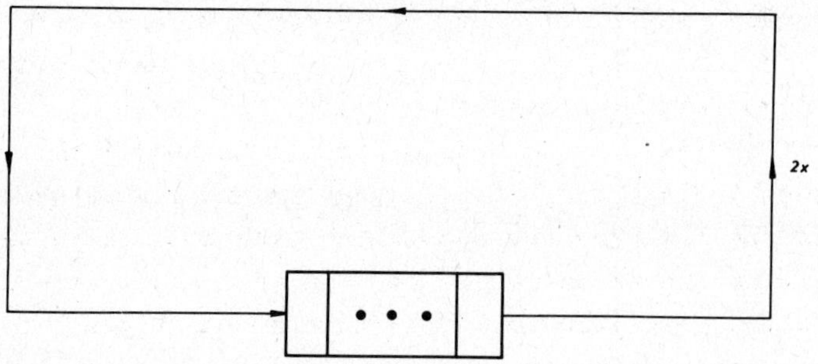

Fig.2.The cell lifespan with k successive subphases

The partial confirmation of the mathematical hypothesis of many successive phases came in 1953 when A.Howard and S.R.Pelc identified in the interphase three different phases,G_1,S,and G_2,in relation with the main observable biochemical event,the duplication of DNA in the cell nucleus. The phases G_1 and G_2 are the "gaps" before and after,respectively,the synthesis. We call this representation the classical scheme of the cell cycle (Fig.3).

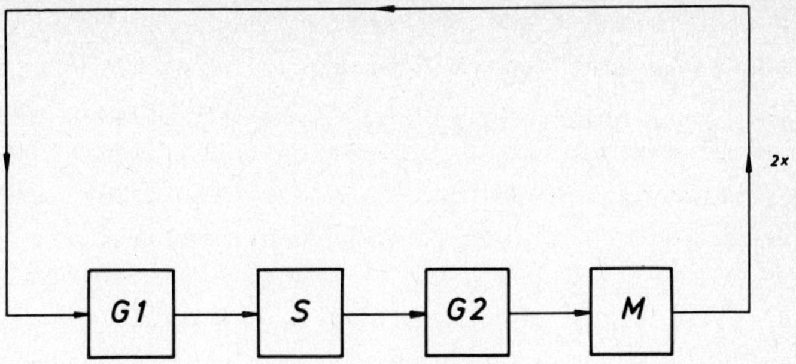

Fig.3.The "classical" scheme of the cell cycle

The cell cycle is an example of those stochastic processes in which the duration of phases are random variables. The appropriate mathematical approaches are the theory of branching processes and the theory of semi-Markov processes. An age-dependent branching process,as Bellman-Harris process (1948,1952),is precisely a stochastic representation of the cell division process (binary splitting),taking into account the lifespan of each cell. In this context,Kendall's model may be interpreted as a 1-dimensional age-dependent branching process or as a multitype Markov branching process. The latter interpretation is mathematically beneficial because a k-phase process with phase durations equally and exponentially distributed is more tractable than a 1-phase process with general distribution.(A generalized multiple-phase birth-and-death process has been presented by M.Takahashi,1968.)

There exist some stochastic models for the classical cell cycle (Weiner,1966;Mode,1971 a;Macdonald,1973),having any common hypotheses. The first one is the independence hypothesis:each cell has a growth pattern independently and identically distributed as the parent cell,and independent of other cells. Secondly,the time spent in phase i,given that the next transition is into phase j is a random variable θ_{ij},$1\leq i<$ $<j\leq 4$,with (nonlattice) pdf $F_{ij}(t)$ dependent on the phases i and j but otherwise independent of the state of the system. H.Weiner (1966) introduced the transition probability matrix $\underline{P}$ which is irreducible and with zero trace. Hence,any phase except the mitotic one may be the initial phase of a newborn cell. C.J.Mode (1971 a,b) introduced the matrix $\underline{M}$ whose elements m_{ij} represent the mean number of cells in phase j produced by a cell in phase i. Because no cell death hypothesis is assumed, clearly $\sum_{j=1}^{4} m_{ij}=1$,$1\leq i\leq 3$, In phase 4 the cell gives birth to two daughter cells which enter phase 1 ($m_{ij}=0$,$2\leq j\leq 4$). The form of pdf for the cell

cycle times is still under discussion. The commonly used distributions are the inverse normal, the lognormal, and the gamma. Macdonald (1973) concluded that the rate of proliferation of the cell population will depend only on m and the pdf of the total cycle time and not otherwise on the phase structure of the cell cycle (add: in its classical form).

But in the last decade this phase structure rapidly changed. In order to explain why some cell populations (e.g. liver, bone marrow stem cells) alter their production rate, H.Quastler (1963) and L.G.Lajtha (1963) assumed the existence of a pre-synthetic quiescent phase, calling it G_o phase, following the accepted "gap" nomenclature. This phase represents a decision stage ("dichophase": Bullough, 1963) or a reservoir from which cells could be randomly triggered in order to supply cells for division when required (see Burns and Tannock, 1970). A second (post-synthetic) quiescent phase with influx from G_2 has also been assumed : S.Gelfant (1963) described G_2 subpopulations having different and specific physiological requirement for mitosis.

A cell population is then inhomogeneous, having at least two subpopulations with dividing and nondividing cells. The fraction of dividing (proliferating) cells has been called the growth fraction (Mendelsohn, 1960) or the proliferating pool (Kisieleski et al., 1961). A new dynamics is imposed: the proliferative "P" cells from the proliferative compartment may leave it going to the nonproliferative compartment as "Q" cells which either retain their proliferative potential or are totally incapable of replication (Cairnie et al., 1965). This transition process is called "cell loss" (Steel, 1967) and is considered a major factor controlling cell growth. It has been shown, for example, that in some cases the rate of cell loss in malignant tumors may be as much as 80% of the rate of tumor cell production (Terz et al., 1971; Cooper et al., 1975). A cell population will be defined by three kinetic parameters: the cell cycle time, the growth fraction, and the rate of cell loss.

The death ("apoptosis": Kerr et al., 1972) hypothesis must also be admitted: this can be age- or phase-dependent, particularly linked with G_2 or M phases (Kerr and Searle, 1972; Cooper, 1973). This process has been particularly studied by P.Jagers (1970), also introducing G_o-phase (1975).

In the recent schemes of the cell cycle two quiescent phases, Q_1 and Q_2, are supposed as well as the death hypothesis. In P.Dombernowsky et al. (1973) scheme no recycling from Q_1 or Q_2 is assumed ; this possibility is present in the scheme proposed by K.B.Woo et al. (1975) with transitions $Q_1 \rightarrow S$ and $Q_2 \rightarrow M$.

The present work is dealing with branching models for new cycle schemata. In Section 2 we introduce a model for a cell cycle including both proliferation and quiescent compartments and also the spontaneous cell death. We call away "cell cycle" the process representing the sequence of events which mark the cell life,even if some cells in the considered cell population never cycle. Our computer simulations show that to a certain extent the behaviour of a cell population depends on its cycle structure (Section 6). The problem of control of a cell cycle is discussed in Section 3 with respect to the recent studies on controlled branching processes (Levina et al.,1968;Fujimagari,1972;Labkovskii, 1972;Sevast'yanov and Zubkov,1974;Zubkov,1974;Lipow,1975).

A comprehensive cell cycle model where each phase may be made up of various stages is described,together with its control problem,in Section 4. In Section 5 we generalize this model for cells of different types.

2.A BRANCHING MODEL FOR A CELL CYCLE WITH PROLIFERATION AND QUIESCENT COMPARTMENTS

Our first model - called Model A - is presented in Fig.4. The proliferation compartment contains the cycling cells and the quiescent compartment contains two quiescent phases,Q_1 and Q_2. The recycling is possible at any time from Q_1 to G_1 and from Q_2 to G_2. The cells in Q_1-phase can spontaneously die.

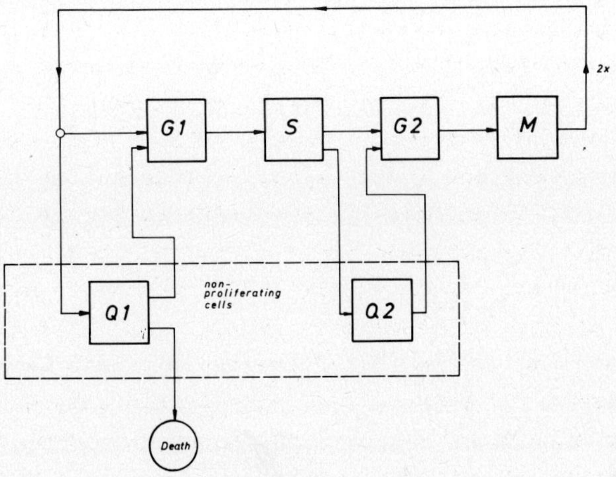

Fig.4.Schematic representation of the cell cycle structure in Model A

The phases are numbered in the following sequence: Q_1, G_1, S, Q_2, G_2, and M, and the following assumptions are made:

A1. Each cell lives and divides independently of each other and of the past.

A2. The individual phase lengths T_i, $1 \le i \le 6$, are independently distributed with gamma distribution Γ_{λ_i, k_i}. We define

$$E[T_i] = \frac{k_i}{\lambda_i}$$

$$D^2[T_i] = \frac{k_i}{\lambda_i^2} = \frac{(E[T_i])^2}{k_i}, \quad k_i \in \{1,2,3,\dots\}.$$

A3. The two daughter cells produced by binary fission at the end of M-phase are going

(1) to Q_1-phase with probability p_{11},

(2) to G_1-phase with probability p_{22}.

The probability of an asymmetric transition (one daughter cell into Q_1 and the other in G_1) is denoted by $2p_{12}$.

A4. The cell in Q_1-phase can either recycle after a random time, entering in G_1-phase with probability $1-\alpha$, or can die with the complementary probability α.

A5. All the cells in G_1-phase move to S-phase.

A6. At the end of S-phase a cell can either leave the cycle entering Q_2-phase with probability β, or run further on in the cycle, entering G_2-phase with probability $1-\beta$.

A7. The cells in Q_2-phase enter G_2-phase and all cells in G_2-phase enter M-phase.

A8. After the mitosis the expected number of cells entering the quiescent phase Q_1 is $m_1 = 2(p_{11}+p_{12})$, and the expected number of cells cycling further on is $m_2 = 2(p_{12}+p_{22})$.

The model we built up has 16 parameters, namely $\lambda_i, k_i (1 \le i \le 6), p_{11}, p_{22}$ $(2p_{12} = 1-p_{11}-p_{22})$, α, and β. This model is equivalent to a multitype continuous time Markov branching process (see Athreya and Ney, 1972, p.199) with $k = k_1 + \dots + k_6$ dimensions. If we denote by $M_{ij}(t)$ the expected number of cells of type j at time t given that the process started at t=0 with one cell of type i, $1 \le i, j \le k$, that is, $M_{ij}(t) = E[Z_j^{(i)}(t)]$, then the mean matrix $\underline{M}(t) = \{M_{ij}(t)\}$, $1 \le i, j \le k$, $t \ge 0$, has by its semigroup property an infinitesimal generator $\underline{A}$ such that

$$\underline{M}(t) = \exp\{\underline{A}t\}.$$

In our model the explicit forms of the components of $\underline{A}$ are the following:

$$A_{ii} = -\lambda_j \;,\; K_{j-1} < i \le K_j \;,\; 1 \le j \le 6 \quad (K_j := \sum_{i=1}^{j} k_i \;, 0 \le j \le 6)$$

$$A_{i,i+1} = \lambda_j \;,\; K_{j-1} < i < K_j \;,\; 1 \le j \le 6$$

$$A_{\delta_j,\delta_j+1} = \begin{cases} \lambda_j \;, & j=2,4,5 \;;\; \delta_j = K_j \\ \lambda_1 \alpha \;, & j=1 \\ \lambda_3 \beta \;, & j=3 \end{cases}$$

$$A_{\delta_j,\delta_h+1} = \begin{cases} \lambda_3(1-\beta) \;, & j=3, h=4 \\ \lambda_6 m_2 \;, & j=6, h=1 \end{cases}$$

$$A_{\delta_j,1} = \lambda_6 m_1 \;,\; j=6$$

$$A_{ij} = 0 \text{ otherwise}$$

For $k_1 = \ldots = k_6 = 1$, $\underline{A}$ has the following form:

$$\underline{A} = \begin{pmatrix} -\lambda_1 & \lambda_1\alpha & 0 & 0 & 0 & 0 \\ 0 & -\lambda_2 & \lambda_2 & 0 & 0 & 0 \\ 0 & 0 & -\lambda_3 & \lambda_3\beta & \lambda_3(1-\beta) & 0 \\ 0 & 0 & 0 & -\lambda_4 & \lambda_4 & 0 \\ 0 & 0 & 0 & 0 & -\lambda_5 & \lambda_5 \\ \lambda_6 m_1 & \lambda_6 m_2 & 0 & 0 & 0 & -\lambda_6 \end{pmatrix}$$

The expected number of cells in the population behaves as $e^{x_o t}$ where x_o is the (real) solution with the greatest real part of the equation

$$f(x) = \det|x\underline{I} - \underline{A}| = 0,$$

where $\underline{I}$ is the identity matrix. We have

$$f(x) = \prod_{i=1}^{6} (x+\lambda_i)^{k_i} - \prod_{i=1}^{6} \lambda_i^{k_i} [\beta + (1-\beta) \cdot \left[\frac{x+\lambda_4}{\lambda_4}\right]^{k_4}] \cdot$$

$$\cdot [(1-\alpha) m_1 + \left[\frac{x+\lambda_1}{\lambda_1}\right]^{k_1} m_2] \,,$$

and

$$f(0) = \prod_{i=1}^{6} \lambda_i^{k_i} (\alpha m_1 - 1) \,,$$

with $f(0) > 0$ iff $\alpha m_1 > 1$, $f(0) = 0$ iff $\alpha m_1 = 1$, and $f(0) < 0$ iff $\alpha m_1 < 1$.

Further considerations lead to the following:

$$x_o>0 \text{ iff } \alpha m_1<1$$

$$x_o=0 \text{ iff } \alpha m_1=1$$

$$x_o<0 \text{ iff } m_1>1.$$

If the process begins with one single cell in phase G_1, the extinction probability q is

$$q=1 \text{ iff } \alpha m_1\geq 1$$

$$q<1 \text{ iff } \alpha m_1<1$$

$$q=0 \text{ iff } \alpha m_1=0.$$

The total number of cells increases exponentially, with probability 1-q. Analogously to all non-singular branching processes, the above considered process is unstable. Such a model describes the growth of a real cell population which starts with few cells and produces a certain amount of cells. But above this level the growth rate diminishes until an equilibrium is possibly reached. It is experimentally confirmed that the relative frequencies of the cell cycle phases as well as the cell lifetime often change (see, e.g., Kauffman, 1968; van den Biggelaar, 1971).

3. THE CONTROL PROBLEM FOR MODEL A

A decisive change in the behaviour of our Model A can be obtained if at least the cell loss parameters m_1 and α are dependent on the total number n of cells. We will consider the following dependence:

$$E[T_1]= c_1[1+f_1(n)],$$

$$\alpha=\frac{\alpha_1+\alpha_2 f_2(n)}{1+f_2(n)},$$

$$\beta=\frac{\beta_1+\beta_2 f_3(n)}{1+f_3(n)},$$

and

$$P=\frac{P^{(1)}+P^{(2)}f_4(n)}{1+f_4(n)}.$$

Here c_1, α_1, β_1, and $P^{(1)}$ are the process parameters when $n\leq n_o$ (n_o is a threshold), and

$$P^{(1)}=\begin{pmatrix} p_{11}^{(1)} & p_{12}^{(1)} \\ p_{12}^{(1)} & p_{22}^{(1)} \end{pmatrix}$$

Consequently, $m_1^{(1)}=2(p_{11}^{(1)}+p_{12}^{(1)})$ and $m_2^{(1)}=2(p_{12}^{(1)}+p_{22}^{(1)})$. The other parameters $\alpha_2(=1), \beta_2$, and $P^{(2)}$ are the "asymptotic" parameters when $f_i(n)$, i=

=2,3,4,unlimitedly increases. Similarly,

$$P^{(2)} = \begin{pmatrix} p_{11}^{(2)} & p_{12}^{(2)} \\ p_{12}^{(2)} & p_{22}^{(2)} \end{pmatrix} ,$$

with $m_1^{(2)}=2(p_{11}^{(2)}+p_{12}^{(2)})$ and $m_2^{(2)}=2(p_{12}^{(2)}+p_{22}^{(2)})$. The function $f_i(n)$ is defined by

$$f_i(n) = \max\{0, g_i(\max(0, \frac{n-n_o}{n_o}))\} , \quad 1 \le i \le 4 ,$$

where $g_i(x)$ are polynomials in x,with $g_i(0)=0$. Then $f_i(n) \geq 0$ for $n>n_o$ and equal to zero for $n \le n_o$. If $\alpha_1 m_1^{(1)} < 1$, $m_1^{(2)} > 1$ and the polynomials g_1 and g_2 monotonely increase for x>0,then there exists a number $n' \geq n_o$ such that

$$\alpha(n)m_1(n) \le 1 \quad , \text{ for } n \le n'$$
$$> 1 \quad , \text{ for } n > n' .$$

Moreover,if $\alpha_1=0$,the process might have a nontrivial stationary distribution. This assumption remains to be proved in the framework of the theory of controlled branching processes.

4.THE GENERALIZED MODEL OF THE CELL CYCLE WITH MULTIPLE STAGES

Biochemical studies evidenced that the separation of the cell lifespan into four or six phases is not so strict; there are good reasons to speak about an "early" or on a "late" X-phase,even on "late X_1-early X_2" phase (Baserga,1968). It is then possible to produce schemata for the cell cycle as that presented in Fig.5.

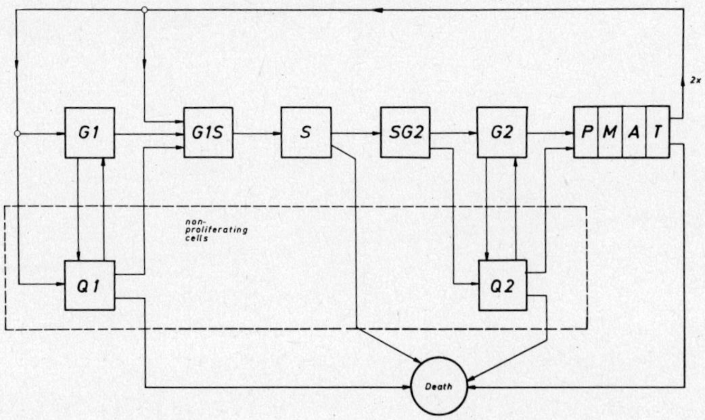

Fig.5.A general scheme for the cell cycle

The experiments allow the hypothesis that each conventional phase (Q_1, $G_1,\dots,M$) is in reality constituted of a group of stages. This hypothesis leads to the construction of a comprehensive branching model for a cell cycle with k+1 phases, each phase having a number s_i, $1\le i\le k+1$, of stages. We introduce the following general assumptions:

B1. Each cell lives and divides independently of each other and of the past.

B2. The length T_i of phase i, $1\le i\le k+1$, is independently and exponentially distributed, with $E[T_i]=1/\lambda_i$.

B3. At the end of phase i, $1\le i\le k$, the cell enters phase j, $0\le j\le k+1$, with probability r_{ij}, the fictitious phase 0 may be interpreted as an absorbing state, the cell death. ($r_{ij}\ge 0$, $r_{ii}=0$, $\sum_{j=0}^{k+1} r_{ij} = 1$).

B4. At the end of phase k+1 a cell either dies with probability q, or gives birth to two daughter cells with probability 1-q.

B5. The probability of an asymmetric division (one daughter cell into phase i and the other into phase j) is p_{ij}, $1\le i,j\le k$ ($p_{ij}=p_{ji}\ge 0$, $\sum_{i,j} p_{ij}=1$).

This model has a general structure and is, clearly, a generalization of Model A introduced in Section 2. With the assumption that the first conventional phase (the phase Q_1 in Model A) contains s_1 different stages, we have for example

$$\lambda_i = \lambda_1 \text{ , for } i\le s_1$$

$$r_{i,i+1} = 1 \text{ , for } i<s_1$$

$$r_{s_1,s_1+1} = 1-\alpha$$

$$r_{s_1,0} = \alpha$$

$$r_{ij} = 0 \text{ , for } i\le s_1 \text{ else}$$

then $E[T_{Q_1}]=s_1/\lambda_1$, $D^2[T_{Q_1}]=s_1/\lambda_1^2$.

The infinitesimal generator $\underline{B} \in \mathbb{R}_{k+1,k+1}$ of the mean matrix of this process has the following components:

$$B_{ii} = -\lambda_i \text{ , } 1\le i\le k+1$$

$$B_{ij} = \lambda_i r_{ij} \text{ , } 1\le i\le k, 1\le j\le k+1, i\ne j$$

$$B_{k+1,j} = 2\,\lambda_{k+1}(1-q) \sum_{i=1}^{k} p_{ij} \text{ , } 1\le j\le k.$$

The behaviour of the considered branching process is determined by the (real) eigenvalue ρ with the greatest real part of matrix $\underline{B}$ if no singular subprocess exists.

In the controlled process all parameters may depend on the total number n of living cells in the following form:

$$Y(n) = \frac{Y_1 + Y_2\ f(n)}{1 + f(n)}$$

with

$$f(n) = \max\{\ 0\ ,\ g[\max(0,\frac{n-n_o}{n_o})]\ \}\ ,$$

where $g(x)$ is a polynomial in x, with $g(0)=0$, and n_o is a threshold for n. This special form of dependence means that Y_1 is the initial value of $Y(n)$ if $n \le n_o$, and Y_2 is the asymptotical limit of $Y(n)$ if $g(x)$ unlimitedly increases. In such a controlled process, where all parameters are functions of n, ρ also depends on n. A sufficient condition for the existence of a non-degenerate limit distribution might be $q(n)=0$ and $r_{i0}(n)=0$, $1 \le i \le k$, for $n \le n_o$, and $\rho(n) \le \varepsilon < 0$ for $n \ge n' > n_o$. The existence of a stationary distribution for a 1-dimensional branching process with population size dependence can be proved (see Lipow, 1975).

5. THE GENERALIZED MODEL WITH SEVERAL CELL TYPES

Let us now suppose that in a cell population there are cells of different types ("colours"). The model we consider is then a k+1 phases / u types branching model, with the additional hypothesis that the change of cell type can occur only in the division phase k+1. The parameters are the following:

$\lambda_i^{(c)}$, $1 \le c \le u$, $1 \le i \le k+1$, the parameter of the exponential distribution of the holding time in phase i for a cell of "colour" c;

$r_{ij}^{(c)}$, $1 \le c \le u$, $1 \le i \le k$, $0 \le j \le k+1$, the probability that a cell of colour c moves from phase i to phase j ($r_{ii}^{(c)}=0$, $\sum_{j=0}^{k+1} r_{ij}^{(c)} = 1$);

$q^{(c)}$, $1 \le c \le u$, the death probability of a cell of colour c at the end of phase k+1;

$p_{c'i,c''j}^{(c)}$, $1 \le c,c',c'' \le u$, $1 \le i,j \le k$, the probability that two daughter cells generated by a parent of colour c have an asymmetric division: one daughter cell is of colour c' and enters phase i, and the other one is of colour c" and enters phase j.

$$(\ p_{c'i,c''j}^{(c)} = p_{c''j,c'i}^{(c)} \ge 0\ ,\ \sum_{\overline{c'i,c''j}} p_{c'i,c''j}^{(c)} = 1\ ).$$

The parameters are the same for the controlled process.

Such a generalized process does not essentially require different theoretical considerations. The analogous matrix $\underline{B}'$ has almost block diagonal form;if it is reducible,this suggests the existence of several subprocesses which may develop independently.

6.COMPUTER SIMULATIONS

We present in this section some simulation experiments realized with a flexible program in APL. This program was run on IBM 370/158 machine of the German Cancer Research Center, Heidelberg.

In order to show the differences of cell growth depending on the cell cycle structure,we simulated our Model A and a variant called Model B. In the latter case,the place of the Q_2-phase is now between G_2- and M-phase;a cell in Q_2-phase can either go into M-phase or die. The probability for this transition as well as the resting time in Q_2-phase are dependent on the number of cells. Both models are controlled in the sense of Section 3. For each model we also realized two versions:in version 1 the length of Q_1-phase in equilibrium state is about 2.6-ply of the length of the proliferation cycle, while in version 2 the expected duration of Q_1-phase is only 1.3-ply of the length of the proliferation cycle. We used for the length of G_1,S and G_2 phases the data reported by Y.Okumura and T.Matsuzawa (1975) for Yoshida sarcoma cells. The other parameters have been chosen according to our experience with similar simulations. (see fig. 6 and 7).

If we compare Model A with Model B, we see that they essentially differ in their behaviour after exceeding the threshold (n_o=250 is the threshold for the 4 presented simulations shown in fig.6 and 7.) up to the approximate equilibrium. The reduction of the growth rate in case of Model A sets in later than in Model B, so that the equilibrium state of Model A is momentarily exceeded. This effect is obvious in the simulations with the data of version 2.

In general, the curves representing the total number of cells as well as the curves of the cells in phases Q_1 und S are distinct. The number of cells in the other phases is at low level, so that in the chosen scale the corresponding curves can no longer be clearly distinguished. The different lengths of the Q_1-phase modify the relative frequencies of cells in the other phases.

A 2-type model whose phase configuration corresponds to the above mentioned configuration of Model B has also been simulated, introducing two types of cells: mammary gland cells and mammary carcinoma cells

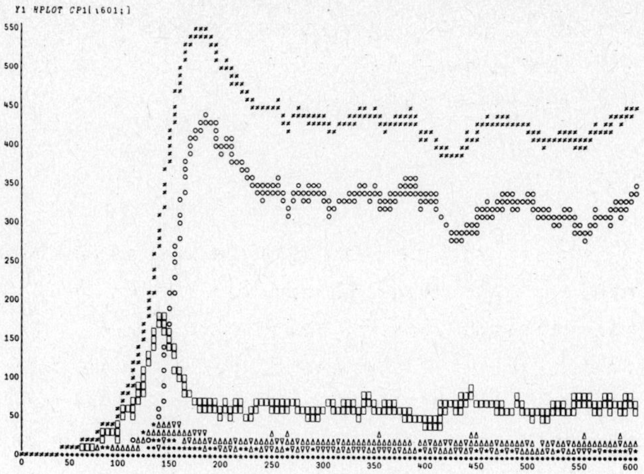

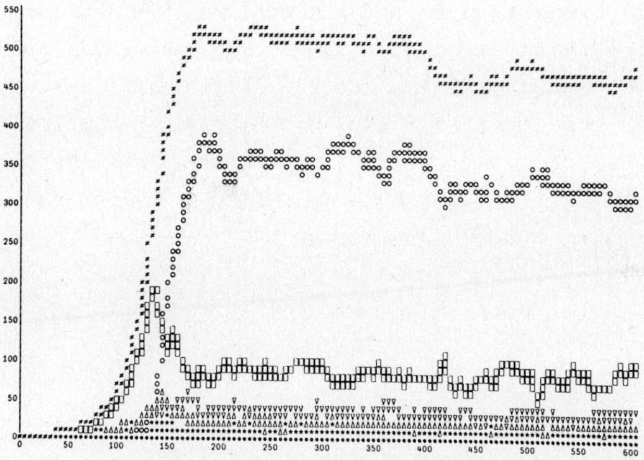

Fig.6. Computer drawings for the simulation of

Model A Version 1

Model B Version 1

≠ :total number of cells, O :Q_1-phase, * :G_1-phase,

▯ :S-phase, ∇ :Q_2-phase, Δ :G_2-phase, ∘ :M-phase

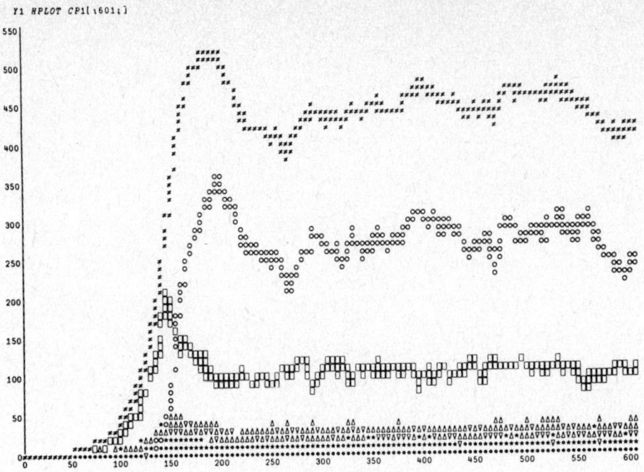

Fig.7. Computer drawings for the simulation of

Model A Version 2

Model B Version 2

≠ :total number of cells, O :Q_1-phase, * :G_1-phase, □ :S-phase, ∇ :Q_2-phase, Δ :G_2-phase, ∘ :M-phase

Fig.8. Computer drawings for the simulation of the 2-type model
≠ : total number of cells, O :normal cells, * :malignant cells

(for the data see Bresciani,1965). In this model, a daughter cell of type 1 (normal) has three alternatives:enter phase Q_1 or phase G_1 without changing its "colour",or enter phase Q_1 as a malignant cell (Type 2). A daughter cell of type 2 cannot change its "colour":it enters either phase Q_1 or phase G_1. The data we had at our disposal referred only to the classical cell cycle;therefore, we chose the other parameters according to our previous simulation experiments and we fixed the thresholds as follows:$n_o^{(1)}=250$,$n_o^{(2)}=300$. The simulation starts with one normal cell in G_1-phase. The probability of colour changing is zero for a population size below its threshold $n_o^{(1)}$,and is about 0.03 after reaching the equilibrium for the normal cells. (see fig. 8).

The effect of the competition of the two cell populations is clearly visible. The tumour cells succeed in reducing the number of the normal cells.

We did not yet simulate models with more than two cell types (e.g. a multi-step model for the carcinogenesis) because of lack of biological data.

REFERENCES

1.Athreya,K.B.,Ney,P.E.(1972).Branching Processes. Berlin-Heidelberg-New York:Springer-Verlag

2.Baserga,R.(1968).Biochemistry of the cell cycle:A review. Cell Tissue Kinet.1,167-191

3.Bellman,R.,Harris,T.E.(1948).On the theory of age-dependent stochastic branching processes. Proc.Nat.Acad.Sci.USA 34,601-604

4.Bellman,R.,Harris,T.E.(1952).On age-dependent binary branching processes. Ann.of Math.55,280-295

5.Bresciani,F.(1965).Acomparison of the cell generative cycle in normal, hyperplastic and neoplastic mammary gland of the C3H mouse. Cellular Radiation Biology,pp.547-557. Baltimore:Williams and Wilkins

6.Bullough,W.S.(1963).Analysis of the life cycle in mammalian cells. Nature 199,859-860

7.Burns,F.J.,Tannock,J.F.(1970).On the existence of a G_o-phase in the cell cycle. Cell Tissue Kinet.3,321-334

8.Cairnie,A.B.,Lamerton,L.F.,Steel,G.G.(1965).Cell proliferation studies in the intestinal epithlium of the rat. Exper.Cell.Res.39,539-553

9.Cooper,E.H.(1973).The biology of cell death in tumours. Cell Tissue Kinet.6,87-95

10.Cooper,E.H.,Bedford,J.,Kenny,T.E.(1975).Cell death in normal and malignant tissues. Advances Cancer Res.21,59-120

11.Dombernowsky,P.,Bichel,P.,Hartmann,N.R.(1973).Cytokinetic analysis of the JB-1 ascites tumour at different stages of growth. Cell Tissue Kinet.6,347-357

12.Fujimagari,T.(1972).Controlled Galton-Watson process and its asymptotic behaviour. 2nd Japan-USSR Symp.on Probability Theory,Vol.2, pp.252-262, Kyoto

13.Gelfant,S.(1962).Initiation of mitosis in relation to the cell division cycle. Exper.Cell.Res.26,395-403

14.Harris,T.E.(1963).The Theory of Branching Processes. Berlin-Göttingen-Heidelberg:Springer-Verlag

15.Howard,A.,Pelc,S.R.(1953).Synthesis of deoxyribonucleic acid in normal and irradiated cells and its relation to chromosome breakage. Heredity 6,Suppl.pp.261-273

16.Iosifescu,M.,Tautu,P.(1973).Stochastic Processes and Applications in Biology and Medicine,Vol.II. Berlin-Heidelberg-New York:Springer-Verlag

17.Jagers,P.(1970).The composition of branching populations:A mathematical result and its application to determine the incidence of death in cell proliferation. Math.Biosci.8, 227-238

18.Jagers,P.(1975).Branching Processes with Biological Applications. London-New York-Sydney-Toronto:Wiley

19.Kauffman,S.L.(1968).Lengthening of the generation cycle during embryonic differentiation of the mouse neural tube. Exper.Cell.Res.49,420-424

20.Kelly,C.D.,Rahn,O.(1932).The growth rate of individual bacterial cells. J.Bacterial.23,147-153

21.Kendall,D.G.(1948).Onthe role of variable generation time in the development of stochastic birth processes. Biometrika 35,316-330

22.Kendall,D.G.(1952).Les processus stochastiques de croissance en biologie. Ann.Inst.H.Poincaré 13,43-108

23.Kerr,J.F.R.,Searle,J.(1972).A suggested explanation for the paradoxically slow growth rate of basal-cell carcinomas that contain numerous mitotic figures. J.Pathol.107,41-44

24.Kerr,J.F.R.,Wyllie,A.H.,Currie,A.R.(1972).Apoptosis:a basic biological phenomenon with wide-ranging implications in tissue kinetics. Brit.J. Cancer 26,239-257

25.Kisieleski,W.E.,Baserga,R.,Lisco,H.(1961).Tritiated thymidine and the study of tumours. Atompraxis 7,81

26.Labkovskii,V.A.(1972).A limit theorem for generalized random branching processes depending on the size of the population. Theor.Probabilities Appl.17,72-85

27.Lajtha,L.G.(1963).On the concept of the cell cycle. J.Cell Comp. Physiol.62,Suppl.,p.143

28.Levina,L.V.,Leontovich,A.M.,Pyatetskii-Shapiro,J.J.(1968).On a regulative branching process. Problems of Information Transmission 4,72-82

29.Lipow,C.(1975).A branching model with population size dependence. Advances in Appl.Probability 7,495-510

30.Macdonald,P.D.M.(1973).On the statistics of cell proliferation. In: The Mathematical Theory of the Dynamics of Biological Populations (R.Hiorns ed.),pp.303-314. London:Academic Press

31.Macdonald,P.D.M.(1974).Stochastic models for cell proliferation. Lecture Notes in Biomath.2,155-161

32.Mendelsohn,M.L.(1960).The growth fraction:a new concept applied to tumours. Science 132,1496

33.Mitchison,J.M.(1973).Differentiation in the cell cycle. In:The Cell Cycle in Development and Differentiation (M.Balls,F.S.Billett eds.), pp.1-11. London:Cambridge Univ.Press

34. Mode,C.J.(1971a).Multitype age-dependent branching process and cell cycle analysis. Math.Biosci.10,177-190

35. Mode,C.J.(1971b).Multitype Branching Processes. New York-London-Amsterdam:Elsevier

36. Okumura,Y.,Matsuzawa,T.(1975).Instability of the duration of G_1 phase of Yoshida sarcoma and ascites hepatomas. Growth 39,331-336

37. Quastler,H.(1963).The analysis of cell population kinetics. In:Cell Proliferation (L.T.Lamerton,R.J.M.Fry eds.),p.18. Oxford:Blackwell

38. Rahn,O.(1932).A chemical explanation of the variability of the growth rate. J.Gen.Physiol.15,257-277

39. Sevast'yanov,B.A.,Zubkov,A.M.(1974).Controlled branching processes. Theor.Probability Appl.19,14-24

40. Takahashi,M.(1968).Theoretical basis for the cell cycle analysis II. J.Theoret.Biol.18,195-209

41. Terz,J.J.,Curutchet,H.P.,Lawrence,W.(1971).Analysis of the cell kinetics of human solid tumours. Cancer 28,1100-1110

42. van den Biggelaar,J.A.M.(1971).Timing of the phases of the cell cycle during the period of asynchronous division up to the 49-cell stage in Lymnaea. J.Embryol.Exp.Morphol.26,367-391

43. Weiner,H.(1966).On age-dependent branching processes. J.Appl.Probability 3,383-402

44. Woo,K.B.,Wiig,K.M.,Brenkus,L.M.(1975).Variation of cell kinetic parameters in relation to the growth rate of the Ehrlich ascites tumour. Cell Tissue Kinet.8,387-390

FORMAL LANGUAGES AS MODELS FOR BIOLOGICAL GROWTH

P.Tautu

Institute for Documentation,
Information and Statistics
German Cancer Research Center
Heidelberg

1. FORMAL BIOLOGICAL SYSTEMS

A formal system is a mathematical structure $S=(V,F,A,\underline{R})$,where

(i). V is a set of symbols,called alphabet, $V=\{a_i|i=0,1,2,\ldots\}$. The set of all finite words (strings) composed with the elements of V is denoted by V^*. Assuming the existence of an empty word Λ,the set V^+ of nonempty words over V is then defined as $V^*-\{\Lambda\}$.

(ii). $F\subset V^*$ is a set of formulae,a language over V.

(iii). $A\subset F$ is a set of initial situations,called axioms.

(iv). $\underline{R}$ is a set of rules of deduction (the primitives). A rule $\rho\in\underline{R}$ is defined as a subset of the Cartesian product $F^n\times F$,where F^n is itself a Cartesian product $F\times\ldots\times F$ having $n\geq1$ factors whose elements are the ordered n-tuples of formulae. Let $F=(x_1,\ldots,x_n)$ be an ordered n-tuple of formulae. If there exists a $\rho\in\underline{R}$ and a formula y such that $F\rho y$ is obtained,then it is said that y is an immediate consequence of the premises $(x_1,\ldots,x_n)$. The formulae in F are the arguments of rule ρ. The set $\underline{R}$ of deduction rules defines the relation of immediate deducibility (see Gross and Lentin,1970). If $F=(x_1,\ldots,x_n)$ are the premises,then a deduction of y is any m-tuple of formulae $F'=(z_1,\ldots,z_m)$ with the property $z_m=y$,if for each i ,$1\leq i\leq m$, one of the following conditions is satisfied:

(a) z_i is an axiom,

(b) z_i is one of the premises,

(c) z_i is immediately deducible from a precedent formula in F'.

If V is finite, F coincides with V^* and $A=\{\alpha\}$,we speak about a <u>combinatorial</u> formal system.

A cellular system can be thought as a formal system if we interpret V as the set of symbols designating cell states (e.g. the cell cycle, genetical and biochemical components,etc.) and the words over V as (linear) assemblies of cells in different states. The axiom is the initial configuration of such a collection of cells. If a word w_1 may be a representation of a cell assembly,then the first modification of the state in a cell produces a new word w_2 by the substitution of a symbol. We say that w_2 is the immediate consequence of w_1 if there is an appropriate deduction rule $\rho \in \underline{R}$. But that modification can be obtained by means of (i) a spontaneous change of the cell state, (ii) a stimulus sent from a neighbouring cell, or (iii) an influence from an environmental factor. If a single deduction rule is used, one may create a sort of ambiguity - which is related with the ambiguity of a word, that is,the number of distinct ways in which it can be generated. The strings representing a cell system are clearly ambiguous.

If we define a formal language L as a subset of V^* for some alphabet V,we figure out that a cell system may have a language for each specific situation,as well as an alphabet. For instance,referring to the cell system described for a carcinogenesis model (Schürger and Tautu, 1976),we conjecture that for each stage there is a particular language with words from a (possibly partially) modified alphabet. As a consequence,the whole biological process is to be represented by a <u>compound</u> language where the axiom in language L_j is the last word of the preceding language L_i.

A grammar G of a language L is defined as the finite set of rules which recursively specifies the words in L. The main problem in the study of formal cell systems is to find the adequate grammar for a particular behaviour,say the normal cell growth. This problem is known as the grammatical (or syntactic) inference problem:"A finite set of symbol strings from some language L and possibly a finite set of strings from the complement of L are known,and a grammar for the language is to be discovered" (Biermann and Feldman,1971). The main task is to devise deductive rules which transform the words in a way consistent with the biological observations. This represents,in fact,the construction of the "hypothesis space" whose motivation must be supported by a profound theoretical work. For example,concepts as polarity,symmetry,apical pattern,branching pattern,etc. are required in order to deduce a grammar explaining the biological development (Herman and Walker,1972).

2. PRELIMINARY DEFINITIONS

The idea to use formal languages as models for the cellular growth originates in the first papers of A.Lindenmayer (1968) where developmental stages of filamentous organisms have been considered as expanding arrays of automata,with or without inputs coming from the neighbouring cells. If a cell is interpreted as a finite automaton changing its states according to a transition ("next-state") function,then,in a natural way, we identify the set of cell states by a finite alphabet and the transition function by a set of some deduction rules. Cell systems as those mentioned above have been called Lindenmayer systems or,simply, L-systems.

Before we define L-systems in terms of formal language theory, let us give the standard definition of a grammar (see,e.g.,Salomaa,1973). A grammar is a system $G=(V_N,V_T,\alpha,\underline{R})$, where V_N and V_T are disjoint finite sets, $V_N \cap V_T=\emptyset$, $V_N \cup V_T=V$,their elements being called <u>variables</u> and, respectively,<u>terminals</u>,with $\alpha\in V_N$ the <u>start variable</u> (or the <u>axiom</u>). $\underline{R}\subset V^+\times V^*$ is a finite nonempty set of ordered pairs (a,b) such that $a\in V^+$ and $b\in V^*$. The elements (a,b) of $\underline{R}$ are called <u>productions</u> (or <u>rewriting rules</u>) and are of the form $a\to b$. The pair $(V,\underline{R})$ is called the <u>general rewriting system</u>.

The binary relation $\to$ ("can be rewritten as") is irreflexive, and if $a\in V^+$,then there are $w_1,w_2,b\in V^*$ such that $w_1aw_2\to w_1bw_2$ (the first axioms in Čulik,1965). Let us denote by W the set of all ordered pairs (x,y), $x,y\in V^*$,such that $x\to y$ in a given grammar G. Then $W=C\underline{R}$,where C is the <u>context operator</u>,which is defined for all binary relations as

$$C\underline{R}=\{(x,y)|\ \exists w_1,w_2\in V^* \text{ and } (a,b)\in\underline{R} \text{ ,such that } x=w_1aw_2 \text{ and } y=w_1bw_2\}.$$

The deductive process ("x derives y") is denoted as $x\Rightarrow y$. We write $x\overset{*}{\Rightarrow}y$ (the reflexive transitive closure of $\Rightarrow$) if there exists a sequence

$$x=w_o,w_1,\ldots,w_k=y \ ,\ k\geq 0$$

such that there are decompositions $w_i=t_ia_iv_i=s_ib_iu_i$ $(a_i,b_i,s_i,t_i,u_i,v_i\in V^*)$ with $(a_i,b_{i+1})\in\underline{R}$, $0\leq i\leq k-1$,and $s_i=t_{i-1}$, $u_i=v_{i-1}$, $1\leq i\leq k$. The sequence $w_o\overset{*}{\Rightarrow}w_k$ is called a derivation from w_o to w_k by G. The set of all derivations is denoted by $\Delta(G)$.

Let $\alpha=(w_o,\ldots,w_k)$ be a derivation and denote by $d\alpha=w_o$ the domain of α and by $c\alpha=w_k$,the codomain. If α and β are two derivations,a new derivation $\beta\circ\alpha$ is obtained as

$$\beta\circ\alpha=(w\underset{\alpha}{\Rightarrow}w'\underset{\beta}{\Rightarrow}w'')$$

if $d\beta=c\alpha$(Walter,1975).

Two derivations are structurally equivalent if they differ only in the order of productions,which cannot ultimately interfere with each other (Buttelmann,1975). Assume the existence of two strings $x,y\in V^*$, $x=w_1a_1w_2a_2w_3$ and $y=w_1b_1w_2b_2w_3$, and also assume two productions in $\underline{R}$, $\rho_1=a_1\to b_1$ and $\rho_2=a_2\to b_2$. Then there are at least two derivations for $x\Rightarrow y$ defined as

$$\Delta_1=(a_1\to b_1\ ,\ w_1)(a_2\to b_2\ ,\ w_1b_1w_2):x\Rightarrow y$$
$$\Delta_2=(a_2\to b_2\ ,\ w_1a_1w_2)(a_1\to b_1\ ,\ w_1):x\Rightarrow y.$$

The difference between Δ_1 and Δ_2 is due entirely to the order of application of ρ_1 and ρ_2 : in Δ_1 we apply the leftmost production,in Δ_2 the rightmost production.

We finally remark that a grammar is a formal system $G=(V_T,F,\alpha,\to)$,in the sense of the definition of Section 1. The language $L\subset V^*$ generated by G is defined as

$$L(G)=\{x\in V_T^*|\alpha\overset{*}{\Rightarrow}x\}.$$

<u>Example</u>(Aho and Ullman,1968). Let G=

({a,b},{u,v,z},α,{α→uαab,α→uab,ba→bc,ua→uv,va→vv,vb→vz,zb→zz}).

$L(G)=\{u^n v^n z^n | n\geq 1\}$. To derive $u^n v^n z^n$ one uses the production α→uαab n-1 times and α⇒uab once to show $\alpha\overset{*}{\Rightarrow}u^n(ab)^n$. Then use the production ba→ab $\frac{1}{2}n(n-1)$ times to show $\alpha\overset{*}{\Rightarrow}u^n a^n b^n$. The remaining four rules then yield $\alpha\overset{*}{\Rightarrow}u^n v^n z^n$. This is a context-sensitive grammar with $|y|\geq|x|$,if x→y.

3. <u>ON L-SYSTEMS AS GENERATIVE DEVICES</u>

In order to define an L-system as a generative device,we must point out that the general rewriting system (V,$\underline{R}$) has a particular biological interpretation. In a biological context,the terminals signify some irreversible states such as death or a nonproliferative terminal differentiation of a living cell. If we intend to represent only systems of proliferating cells,the alphabet is not partitioned into V_N and V_T. Such systems are rightly called <u>developmental</u>. The term <u>propagating system</u> designates a cell system where for every production $(a,x)\in\underline{R}$, $x\neq\Lambda$.

Furthermore,if we acknowledge that in a developmental system the change of cell states occurs simultaneously everywhere,then the productions are to be simultaneously applied to all symbols in each array. This property is somewhat similar with the production application in an unordered scattered context grammar (Greibach and Hopcroft,1969; see Mayer,1972).

The absence of terminals and the synchronous application of rewriting rules are the two principal differences between L-systems and the generative grammars. The language generated by a L-system $L=(V,\alpha,\underline{R})$ is defined as

$$L(L)=\{x\in V^*|\alpha\overset{*}{\Rightarrow}x\}.$$

<u>Example</u>(Rozenberg and Lindenmayer,1973). Consider the L-system

({a,b,c,d,e,f},a,{a→bc,b→df,c→a,d→e,e→e,f→f}).

The system is a representation of the growth of a filamentous organism with apical dividing cells (types a and b),two types of differentiated cells (e,f),and two intermediary cell types (c,d). We have the following sequences

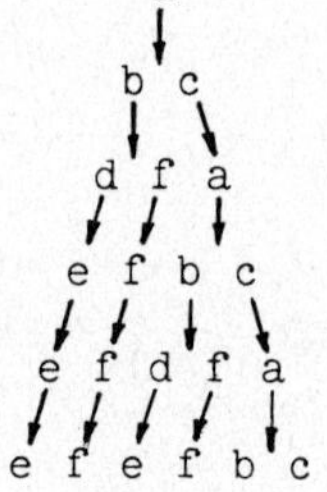

etc.

In the last sequence we remark a banded pattern efef... which may be interpreted as a zonation or as an alternation of substructures. Every string appears at the right segment of the string produced two steps later (the periodicity of a and bc ,for instance). We say that this system generates a right-recurrent sequence with delay two.

If we introduce terminals in an L-system,we obtain the canonical extensions of these systems;if the productions are not synchronously applied,we construct a pure grammar $G=(V,\alpha,\underline{R})$. The language generated by a pure grammar consists of all words over V such that $\alpha_o \overset{*}{\Longrightarrow} w$, $\alpha_o \in \alpha$, if $\alpha \in V^*$. The family of languages $L(L)$ displays a certain resistance to some algebraic operations which characterize the abstract families of languages (see Herman and Rozenberg,1975,pp.188-196).

In the following we consider an L-system in which the cells interact between them and the growth depends on some environmental conditions. Because the system is one-dimensional,the interaction takes place on the left (right) sides of a symbol. We speak about an (m,n)L-system if the production is dependent on the m≥1 left sybols and on the n≥1 right symbols,i.e. the state of a cell depends on its m left neighbours and on its n right neighbours. The productions are environment-dependent and we assume that at each epoch only one subset of productions (called "table") governs the development. This formal construction is motivated by three hypotheses:(1) the interaction between cells,with the existence of neighbourhood conditions,(2) the substitution of deduction rules at each stage of the development, and (3) the control of cell growth by environmental factors.

An (m,n)L-system is defined as a 4-tuple $(V,\underline{R},\alpha,\mu)$,where

(i) V is the finite alphabet,

(ii) $\alpha \in V^+$ is the initial configuration,

(iii) $\mu \notin V$ is the end marker,the environment symbol,

(iv) $\underline{R}=\{R_1,\dots,R_u\}$, u≥1 ,is the finite set of tables of productions:

$$R \subset \bigcup_{\substack{i,j,k,l \geq 0 \\ i+j=m \\ k+l=n}} \{\mu^i\}V^j \times V \times V^k\{\mu^l\} \times V^* ,$$

where for every (w_1,a,w_2) in $\bigcup_{\substack{i+j=m \\ k+l=n}} \{\mu^i\}V^j \times V \times V^k\{\mu^l\}$ there exists an $x \in V^*$ such that $(w_1,a,w_2,x) \in \underline{R}$ (Rozenberg,1974).

The above L-system is denoted as TIL-system (T for "table" and I for "interaction",where I=0 for no interaction). We have four variants depending on the values of T and I:

	$\|R\|>1$	$\|R\|=1$
$m\geq 1, n\geq 1$	T(m,n)L	(m,n)L
m=n=0	TOL	OL

Recently, a new class of grammars, called tally grammars, has been introduced (Savitch, 1975). The languages generated by these grammars include all the L-languages (the main theorem), but also include other languages.

4. ON MOSAIC GRAMMARS

From the discussion above it is clear that in order to realize appropriate grammars for cell systems we must introduce restrictions both on form and on use of productions. Recent papers on L-systems deal with the introduction of different control devices (Ginsburg and Rozenberg, 1975; Rozenberg, 1975; Nielsen, 1975; Asveld, 1975).

The construction of developmental languages gave a new impulse to the research in formal language theory and created a new interdisciplinary field. Knowing the limits of these languages, one may admit that the L-systems theory has at least three functions (Doucet, 1975):

(i) to construct formal models of actual individuals or species and to predict the form of their development;

(ii) to deduce general statements about the general rewriting systems, as for example, "it is impossible to generate this structure without cellular interaction" (strong statement on R) or "it is not necessary to have polar cells in order to generate polar organisms" (weak statement on V);

(iii) to stimulate the research in theoretical biology, by using and creating new theoretical concepts (e.g. internal and external polarity, locally catenative structure, etc.).

One of the greatest impediments in the large application of these grammars is the difficulty to represent 2-or 3-dimensional cell systems (Carlyle et al., 1974; Paz, 1975). The most admitted approach in order to solve the dimensionality problem is the branching interpretation of a string (see Honda, 1971; Hogeweg and Hasper, 1974). For instance, Čulik and Lindenmayer (1974) represented 3-dimensional filamentous organisms as digraphs whose vertices are interpreted as the cells and the edges as a connection between cells. A digraph is simultaneously substituted for every vertex in the initial digraph; then new edges are introduced between the vertices of this new digraph, according to a finite set of special connection rules. The model is basically context-free, with possible context-sensitive extensions. "Structured" OL-systems have been conceived with the aid of labelled trees (Čulik, 1974).

This structural approach led to the idea of graph-L-systems (Nagl, 1975). The difference between this type of graphs and the "webs" (Pfaltz and Rosenfeld,1969;Montanari,1970) is that in the latter the productions are sequential. Moreover,web models cannot indicate the relative positions of the cell neighbours (Mayoh,1974).

Carlyle et al. (1974) constructed a planar model in which an organism is represented as a planar map with cells as labelled countries. A new map is obtained from an already generated map in a context-sensitive way,by simultaneous and binary splitting of every country (cell). The neighbourhood problem is solved by imposing a change of label of each cell if the number of its neighbours exceeds a certain threshold.

The approach we suggest is the construction of a mosaic grammar (see Ohta,1975) in a near ressemblance with the tessellation generative device (see Richardson,1972). Let us consider a 2-dimensional structure with uniform squares and call it mosaic. Associate to each square a symbol from an alphabet V ,operation which realizes a configuration ξ over the mosaic M. We say that square x contains the symbol a in ξ if $\xi(x)=a$. Because of the geometry,the number of neighbours remains constant and a neighbourhood condition specifies the appropriate production. An elaborated discussion will be presented in a forthcoming paper.

REFERENCES

1. Aho,A.V.,Ullman,J.D.(1968).The theory of languages. Math.Systems Theory 2,97-125
2. Asveld,P.R.J.(1975).On some controlled ETOL-systems and languages (abstract).Conf.on Formal Languages,Automata and Development (Noordwijkerhout,The Netherlands)
3. Biermann,A.,Feldman,J.(1971).A survey of grammatical inference.In: Frontiers of Pattern Recognition(S.Watanabe,ed.).New York:Academic Press
4. Buttelmann,H.W.(1975).On the syntactic structures of unrestricted grammars.I.Information and Control 29,29-80
5. Carlyle,J.W.,Greibach,S.,Paz,A.(1974).A two-dimensional generating system modelling growth by binary cell division.IEEE Conf.Record, 15th Annual Symp.on Switching and Automata Theory,pp.1-12
6. Čulik,K.(1965).Axiomatic system for phrase structure grammars. Information and Control 8,493-502
7. Čulik,K.(1974).Structured OL-systems.Lecture Notes in Computer Sci. 15,pp.216-229
8. Čulik,K.,Lindenmayer,A.(1974).Parallel rewriting on graphs and multidimensional development.Acta Informat.(to appear)
9. Doucet,P.G.(1975).On the applicability of L-systems in developmental biology(abstract).Conf.on Formal Languages,Automata and Development (Noordwijkerhout,The Netherlands)
10. Feldman,J.(1972).Some decidability results on grammatical inference and complexity.Information and Control 20,244-262
11. Ginsburg,S.,Rozenberg,G.(1975).TOL schemes and control sets. Information and Control 27,109-125
12. Greibach,S.,Hopcroft,J.(1969).Scattered context grammars. J.Comput.System Sci.3,323-347

13. Griffiths,T.V.(1968).Some remarks on derivations in general rewriting systems.Information and Control 12,27-54
14. Gross,M.,Lentin,A.(1970).Introduction to Formal Grammars.Berlin-Heidelberg-New York:Springer-Verlag
15. Herman,G.T.,Rozenberg,G.(1975).Developmental Systems and Languages. Amsterdam:North-Holland
16. Herman,G.T.,Walker,A.D.(1972).The syntactic inference problem applied to biological systems.In:Machine Intelligence 7,pp.341-356. Edinburgh:Edinburgh University Press
17. Hogeweg,P.,Hasper,B.(1974).A model study of biomorphological description.Pattern Recognition 6,165-179
18. Honda,H.(1971).Description of the form of trees by parameters of the tree-like body:effects of the branching angles and the branch length on the shape of the tree-like body.J.Theoret.Biol.31,331
19. Lindenmayer,A.(1968).Mathematical models for cellular interactions in development.I,II.J.Theoret.Biol.18,280-315.
20. Mayer,O.(1972).Some restrictive devices for context-free grammars. Information and Control 20,69-92
21. Mayoh,B.H.(1974).Multidimensional Lindenmayer organisms. Lecture Notes in Computer Sci.15,pp.302-326
22. Montanari,U.G.(1970).Separable graphs,planar graphs and web grammars.Information and Control 16,243-267
23. Nagl,M.(1975).Graph-Lindenmayer-systems and languages.Arbeitsberichte Inst.Math.Maschinen 8,16-63
24. Nielsen,M.(1975).EOL systems with control devices.Acta Informat. 4,373-386
25. Ohta,P.A.(1975).Mosaic grammars.Pattern Recognition 7,61-65
26. Paz,A.(1975).Multidimensional parallel-rewriting generating systems (abstract).Conf.on Formal Languages,Automata and Development (Noordwijkerhout,The Netherlands)
27. Penttonen,M.(1974).One-sided and two-sided context in formal grammars.Information and Control 25,371-392
28. Pfaltz,J.L.,Rosenfeld,A.(1969).Web grammars.Proc.Intern.Joint Conf. on Artificial Intelligence,pp.609-619
29. Richardson,D.(1972).Tessellations with local transformations. J.Comput.System Sci.6,373-388
30. Rozenberg,G.(1974).Theory of L-systems from the point of view of formal language theory.Lecture Notes in Computer Sci.15,pp.1-23
31. Rozenberg,G.(1975).On OL-systems with restricted use of productions.J.Comput.System Sci.(to appear)
32. Rozenberg,G.,Lindenmayer,A.(1973).Developmental systems with locally catenative formulas.Acta Informat.2,214-248
33. Salomaa,A.(1973).Formal Languages.New York:Academic Press
34. Savitch,W.J.(1975).Some characterizations of Lindenmayer systems in terms of Chomsky-type grammars and stack machines.Information and Control 27,37-60
35. Schürger,K.,Tautu,P.(1976).A Markovian configuration model for carcinogenesis.This volume
36. Smith III,A.R.(1971).Two-dimensional formal languages and pattern recognition by cellular automata.IEEE Conf.Record,12th Annual Symp. on Switching and Automata Theory,pp.144-152
37. Walter,H.(1975).Topologies on formal languages.Math.Systems Theory 9,142-158
38. Walters,D.A.(1970).Deterministic context-sensitive languages.I,II. Information and Control 17,14-61

GRAPH REWRITING SYSTEMS AND THEIR APPLICATION IN BIOLOGY

Manfred Nagl
Institut f. Mathematische Maschinen u. Datenverarbeitung (II)
Universität Erlangen-Nürnberg

1. INTRODUCTION

In sequential rewriting systems on strings, which are called Chomsky-grammars, derivation is defined as the replacement of a substring by another one according to a rewriting rule. The rest of the host string remains unchanged. Since their introduction in 1959, a broad theory has been developped (see e.g. [34]). Starting with [28] and [35], in the last years several authors have generalized this concept to get more or less complex rewriting systems, named graph or web grammars[1],[2],[1o],[15],[21],[22],[23],[27],[29],[31],[36], and[38]. In keeping up the idea of sequential rewriting, in one derivation step on graphs only one subgraph is replaced by another one while the rest of the host graph remains unchanged (cf. Fig. 1). In each derivation step exactly one rewriting rule is applied. This rule has to specify which subgraph is to be replaced (left hand side), which subgraph has to be inserted

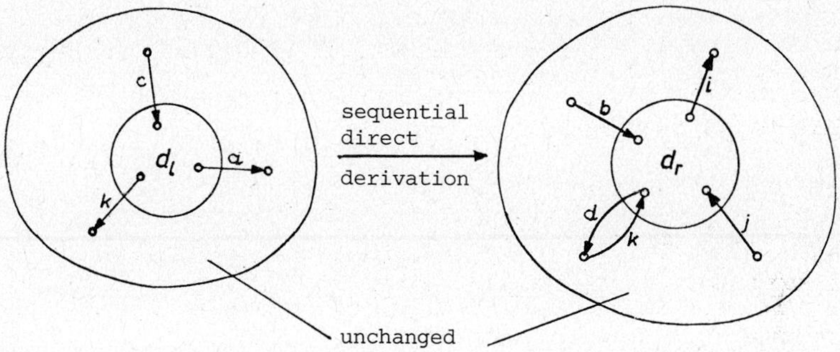

Fig. 1

(right hand side) and, furthermore, how the embedding (incoming and outgoing edges) of the left hand side is transformed if the right hand side is substituted for it. Thus a graph production for sequential rewriting is a triple

$$p = (d_l, d_r, E)$$

with d_l and d_r being graphs, namely the left and right hand side of p respectively, and E being any algorithmic specification for the transformation of the embedding. All approaches mentioned above differ mainly in the way the embedding transformation E is defined and which embedding manipulations the definition allows.

In 1968 Lindenmayer[17] introduced parallel rewriting systems on strings named L-systems which have become a field of enormous interest within the last years (cf. [16],[32],[33]). In such systems, within a derivation step all symbols of a string are replaced simultaneously. Similarly, in parallel rewriting on graphs [6],[7],[9],[11],[2o],[24],[25] the whole graph is rewritten in one derivation step. If the graph d' is the direct derivation of d then we have a subgraph partition of d and d' respectively such that for any subgraph d_l^j of d there is a corresponding d_r^j of d' and vice versa.

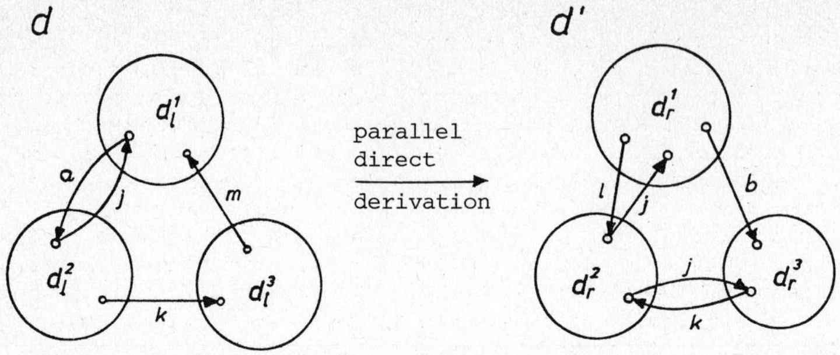

Fig. 2

Any pair (d_l^j, d_r^j) corresponds to a rewriting rule with d_l^j and d_r^j being left and right hand side of the rule respectively. Furthermore, in parallel rewriting we need an algorithmic specification how the connection between left hand sides is transformed to get the connection between the corresponding right hand sides after rewriting. So, in the same way as in the sequential case, a graph production for parallel replacement consists of three parts

$$p=(d_l, d_r, C)$$

where d_l and d_r is the left and right hand side of the rule and C is called the connection transformation.

According to the biological applications we have in mind, we confine us in the following paper to the parallel rewriting of one-node subgraphs (cf. Fig. 3). The resulting rewriting systems we call graph Lindenmayer-systems. Furthermore, (according to [6]) we call left and right hand sides of productions mother nodes and daughter graphs respectively. It should be mentioned, however, that all definitions given in the sequel remain valid for the more general case where more complex subgraphs than single nodes are replaced. As indicated in Fig. 3 the connection structure

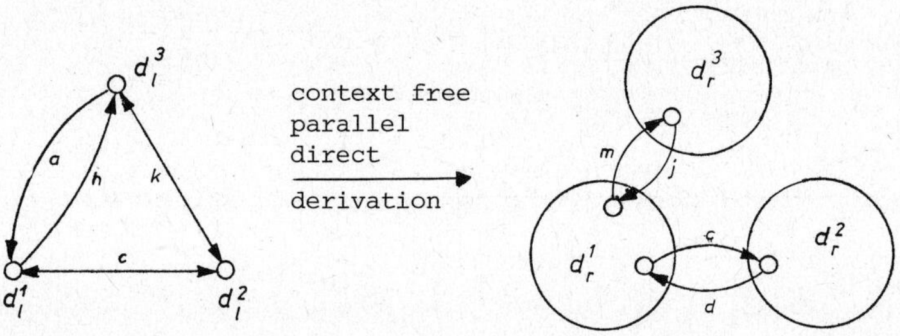

Fig. 3

between the inserted daughter graphs need not be similar to the connection structure between the corresponding mother nodes.

In common Chomsky-grammars and L-systems we use the same definition of a production (besides a restriction in the latter) but different definitions of direct derivation. Similarly, it would be useful (e.g. for the implementation of graph rewriting) if we could use the same definition of a graph production for both rewriting mechanisms on graphs, i.e. if the third component of a graph production could be interpreted as embedding transformation in the sequential case, and as connection transformation in the parallel case as well. This analogy holds true in the following approach: For graph L-systems we only use a restricted form of graph productions defined for graph grammars, i.e. we use context free productions (possibly with deletion) in the sense of [22],[23].

Common L-systems have been introduced to model the development of filamentous plants. They operate on strings, i.e. the description of the shape of an organism is coded within a string. Conversely, any string generated by a L-system must be decoded to get an idea of the form of the corresponding organism. For simple organisms which have linear or tree-like structure, coding in form of strings over an alphabet seems to be reasonable. For more complex organisms this coding is inadequate, because some of the neighbourhood relations are veiled by linearization. To be able to model even the growth of these organisms we propose a coding of the shape of an organism in the form of a graph with labelled nodes and edges. The nodes correspond to the cells as atomic units of the organism and different types of cells are characterized by different node labels. The labelled edges correspond to biological, chemical, geometric, or any other relations between the cells. These graphs more directly reflect the shape of an organism than a string does. So, the process of coding and decoding is a rather trivial task. As already mentioned above,we specialize to one-node graph replacement, i.e. in a direct derivation step each node of a graph is replaced by a daughter graph, according to a finite set of replacement rules. Starting from an axiom graph,all graphs which can be derived in one or more steps belong to the language of a graph L-system.

In section 2 we present a biological example which motivates the introduction of graph L-systems. In section 3 we give the definitions for graph L-systems, table graph L-systems, and extended graph L-systems and the corresponding languages of one of the possible approaches [24],[25]. As this paper is an overview paper , in section 4 we quote some results on graph L-systems and graph grammars, and make some comments on these results but we do not prove them here. In section 5 we present another approach for parallel rewriting on graphs, which is a simplified form of [6] and make some remarks on the relation between these two different parallel mechanisms of section 3 and 5. Finally, in section 6 we sketch some other applications of graph rewriting systems in some fields of computer science.

2. MOTIVATION

To give a motivating example for graph L-systems let us regard the leaf development of the moss Phascum Cuspidatum (cf [18]):

In this moss there are three types of cells: primary, secondary, and tertiary cells, denoted in the following figures by the corresponding starting letters p,s, and t. In each step of development a primary cell gives rise to another primary cell and a secondary cell. A secondary cell can produce a tertiary cell and a secondary cell by a periclinal cell division (division plane parallel to the outline of the leaf) or it may split into two secondary cells by an anticlinal division (division plane perpendicular to the outline). In Fig. 4 the first steps of development are drafted [1]:

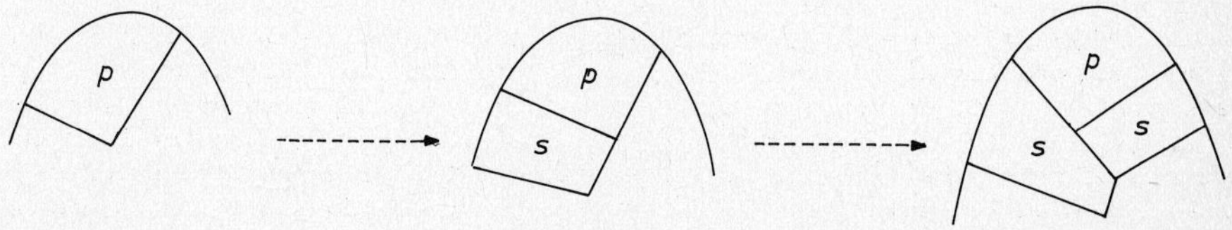

Fig.4.a Fig.4.b Fig.4.c

Fig.4

1) The author is intebted to Prof. Haustein for making some remarks on the development of Phascum leaf.

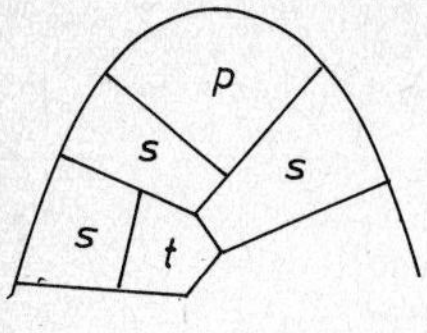

Fig.4.d

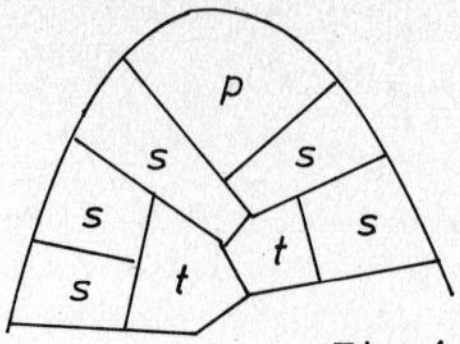

Fig.4.e

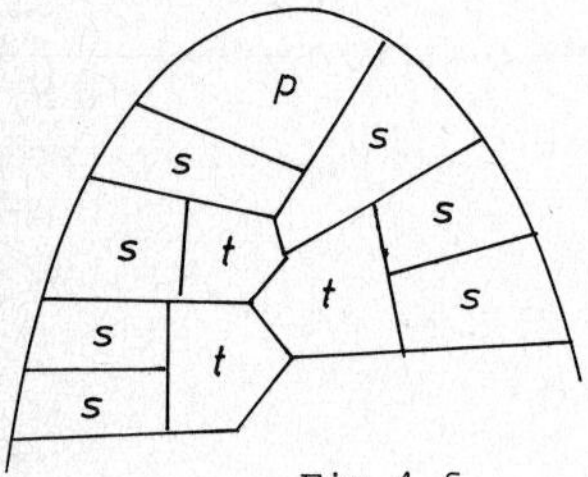

Fig.4.f

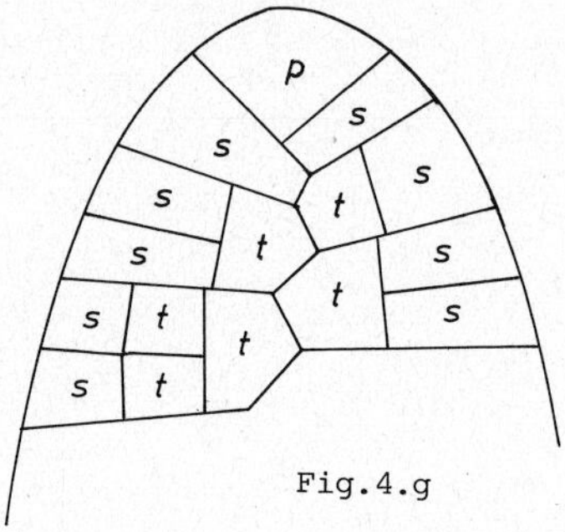

Fig.4.g

Fig. 4

In [18]a nondeterministic OL-system is presented to describe the development of this leaf, i.e. the above pictures are coded by strings. Conversely, the words of the language of the string system have to be decoded to get the corresponding figures. The conventions for this process of coding and decoding are rather complicated: Parantheses are used to indicate imaginary branches within the leaf in alternating left and right positions along the midrib. Brackets of depth two are used to determine adjacent branches within the alternating branches (on the same side of the midrib). The last picture of Fig. 4 for example is coded by the following string

(t(ts)(ts))(t(s)(s))(t(s)(s))(ts)(s)(s)p.

In graph L-systems the above pictures are coded by labelled graphs. The nodes correspond to the cells the different types of which are indicated by different node labels, in our example by p, s, and t. The labelled edges correspond to geometric relations between the cells. In the following example labelled edges, for the reason of simplicity, are expressed by different types of edges, namely

- for the relation to be consecutive cells on the midrib
- for the relation to be consecutive cells on a branch originating in the midrib
- for the connectedness of peripheral cells (direction to the tip of the leaf).

Thus, the picture of Fig. 4.g is coded by graph of Fig. 5. To receive a deterministic graph system we split the label s which stands for "secondary cell" into six substates s, s', s'', s''', s^{IV}, and s^{V}. The new label s stands for a secondary cell after generation from a tip cell, s'' for a secondary cell after the first periclinal cell division, s''' for a secondary cell after the following anticlinal cell division and, finally, s^{V} for a secondary cell after the following second periclinal cell division. The missing states s' and s'^{V} are successors of s and s''' respectively and indicate a reproduction

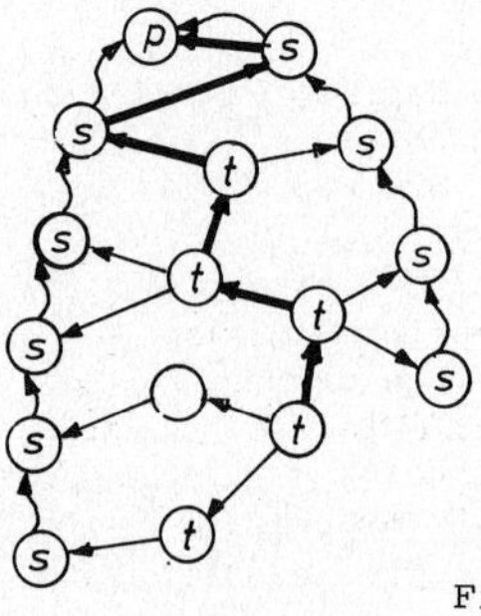

Fig. 5

of the cell without a change of its type.
A graph L-system describing the development of the Phascum leaf is given by the
axiom graph (p)¹
and the
following graph productions, where left of the Symbol "::=" we find the mother node
which is to be replaced, and right of this symbol the daughter graph which is inserted for this mother node:

                                                         connection transformations

p_1: (p)¹ ::= (s)² ⟶ (p)¹ $1\_ = (L\_(1);2)$
                                     $1\frown = (sL\frown(1);1) \cup (s'L\frown(1);2)$

p_2: (s)¹ ::= (s')¹ $1\_ = (L\_(1);1)$ $r\_ = (1;R\_(1))$
                        $1\frown = (L\frown(1);1)$ $r\frown = (1;R\frown(1))$

p_3: (s')¹ ::= (t)² ⟶ (s'')¹ $1\_ = (L\_(1);2)$ $r\_ = (2;R\_(1))$
                                    $1\frown = (L\frown(1);1)$ $r\frown = (1;R\frown(1))$

p_4: (s'')¹ ::= (s''')¹ ⟵ (s''')² $1\_ = (L\_(1);1,2)$
                              $1\frown = (L\frown(1);2)$ $r\frown = (1;R\frown(1))$

p_5: (s''')¹ ::= (s^{IV})¹ $1\_ = (L\_(1);1)$
                        $1\frown = (L\frown(1);1)$ $r\frown = (1;R\frown(1))$

p_6: (s^{IV})¹ ::= (t)² ⟶ (s^{V})¹ $1\_ = (L\_(1);2)$
                             $1\frown = (L\frown(1);1)$ $r\frown = (1;R\frown(1))$

p_7: (t)¹ ::= (t)¹ $1\_ = (L\_(1);1)$ $r\_ = (1;R\_(1))$
                  $1\_ = (L\_(1);1)$ $r\_ = (1;R\_(1))$

p_8: (s^{V})¹ ::= (s^{V})¹ $1\_ = (L\_(1);1)$
                    $1\frown = (L\frown(1);1)$ $r\frown = (1;R\frown(1))$

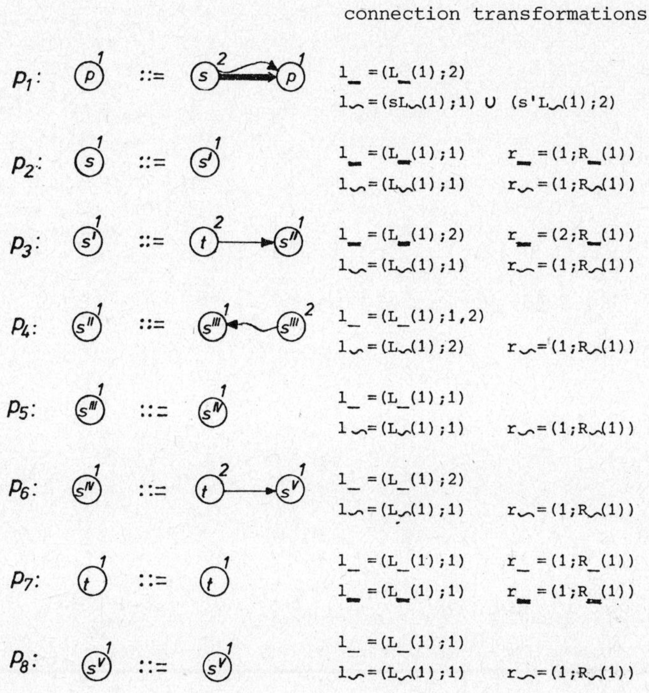

Fig. 6

To understand how this graph L-system works, let us regard the first two steps of
derivation. In the first step the axiom graph consisting of a single node is replaced by a two-node daughter graph using production p_1. In this case there is no problem how to connect daughter graphs as only one daughter graph exists. Therefore, in
the first step the connection transformation of p_1 becomes irrelevant. In the second
step the p-labelled node is rewritten by the same daughter graph using production p_1
again and the s-labelled node is replaced by the daughter node with label s' using
production p_2. In this step we have to generate connections between daughter graphs
by interpreting the connection transformation of the two applied rules. To facilitate
understanding, we give an intermediate graph where the connections between daughter
graphs consists of half-edges(Fig.7.b).These half-edges act only as an aid for understanding and are not really generated.The interpretation of the connection transformation is as follows: Each connection transformation consists of connection components,in
our example $1\blacksquare, 1-, 1\frown$, and $r\blacksquare, r-, r\frown$. The connection component $1\blacksquare$ for example
is responsible for the solid plotted incoming edges, $r\frown$ for the curved outgoing
edges of the corresponding daughter graph. The component $1\blacksquare=(L\blacksquare(1);2)$ is to be
read: Solid plotted incoming edges have to go into node 2 of the daughter graph.
Furthermore, the expression $L\blacksquare(1)$ in $1\blacksquare$ specifies where the edges are coming
from. This expression is to be applied to the graph before the last derivation step,
i.e. to Fig. 7.a,as the node denotation left of the semicolon belongs to the mother

node. It must be read as: Go from the mother node (with label p) along a solid plotted incoming edge. The result consists of the node with label s. The corresponding daughter graph (which is the s'-labelled node) is the origin of an edge which terminates in node 2 of the daughter graph of p_1. As usually a daughter graph consists of more than one node, the component $l_{—}$ only determines a half-edge, in our case half-edge *1* of Fig.7.b. Analogously, $l_{\frown}$ determines: Incoming curved edges are to go into node 1 of the daughter graph if they come from a daughter graph determined by $sL_{\frown}(1)$, and they are to go into node 2 if they come from a daughter graph determined by $s'L_{\frown}(1)$. The expression $sL_{\frown}(1)$ is to interpret: Starting from the p-labelled node, go along an incoming curved edge and take the part of the resulting nodes which are labelled with s. This yields the s-labelled node of Fig.7.a and therefore half-edge *2* is determined. The second part $(s'L_{\frown}(1);2)$ of $l_{\frown}$ becomes irrelevant as $s'L_{\frown}(1)$ does not yield a result in Fig.7.a. The components $r_{—}$ and $r_{\frown}$ of p_2 yield half-edges *3* and *4* respectively. The interpretation is analogous with the difference that the nodes from which the edges originate are written here left of the semicolon, and the daughter graphs where these edges terminate are given by the expressions right of the semicolon. An edge between two daughter graphs is generated iff two half-edges fit together. Thus, by deleting the intermediate graph, after the second derivation step we get Fig.7.c. If the reader knows that the component $(L_{—}(1);1,2)$ of p_4 is only an abbreviation for $(L_{—}(1);1) \cup (L_{—}(1);2)$ then he can follow the next derivation steps given by Fig.7.d-7.g.

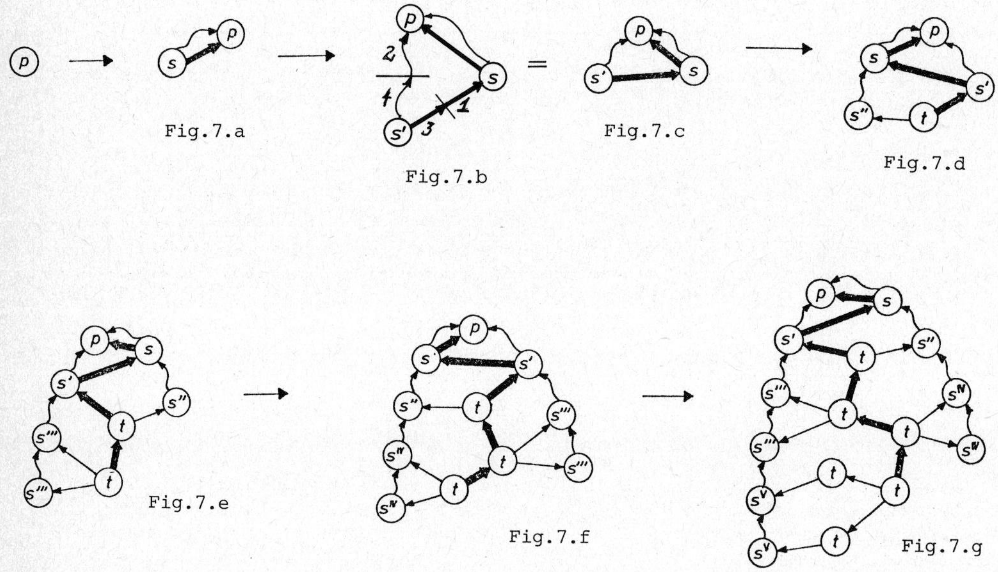

Fig.7.a Fig.7.b Fig.7.c Fig.7.d Fig.7.e Fig.7.f Fig.7.g

Fig.7

The reader may have the legitimate impression that the notation of connection transformations is too overcharged for the description of such simple developments. The reason is that the above notation was introduced to cover more complex examples (cf. Ex. 3.26').

Comparing common L-systems on strings and graph L-systems we can state:

- The derivation mechanism informally introduced above and formally by section 3 is much more complicated than the rewriting mechanism in common L-systems.
- On the other hand graphs directly represent all relations between the cells of an organism. The advantage of graph modeling is that the transformation between a

figure of an organism and its coding, and vice versa, is trivial if we take labelled graphs.

- Common L-systems on strings seem to be applicable only for organisms which have a linear or tree-like structure while graph L-systems are able to describe the development of wider classes of organisms built up by regular rules.

3. DEFINITION OF GRAPH L-SYSTEMS AND LANGUAGES

The underlying structures for our generalized L-systems are not strings over an alphabet but graphs with labelled nodes and edges. We assume Σ_V, Σ_E to be two finite alphabets, the alphabet of node (vertice) labels and edge labels respectively.

Def.3.1: A labelled graph over the alphabets Σ_V, Σ_E is a tuple $d=(K, (\varrho_a)_{a\in\Sigma_E}, \beta)$ where

- . K is a finite set, the set of nodes,
- . ϱ_a is a relation over K for any $a\in\Sigma_E$, i.e. $\varrho_a \subseteq K\times K$,
- . β: $K\rightarrow\Sigma_V$ is a function, called the labelling function.

Any pair $(k_1,k_2)\in\varrho_a$ can be understood as a directed edge from k_1 to k_2 with label a and, furthermore, each node k is labelled with a $\beta(k)\in\Sigma_V$. For denoting the set of nodes we usually use a subset of $\mathbb{N}_0$ [2)].

A labelled graph, shortly l-graph, can be used to describe geometric, biological, physical, technical etc. objects: The nodes represent substructures which are differentiated by their node labels and the relations ϱ_a are used to express geometric, biological, physical, or any other actually interesting relations between these substructures. By going from a class of objects to a corresponding set of l-graphs, any ϱ_a corresponds to a special relation between the parts of the objects. The number of relations, i.e. the cardinality of Σ_E depends on the application. Trivially, a l-graph over Σ_V, Σ_E can be extended to a l-graph over Σ_V, Σ_E' with $\Sigma_E \subseteq \Sigma_E'$, by adding empty relations.

Any word $w=x_1\ldots x_m$ over an alphabet V can be described as a l-graph over V,{a}, namely $d_w=(K,\varrho_a,\beta)$ with $K=\{1,\ldots m\}$, $\varrho_a=\{(i,i+1) \mid 1\leq i\leq m-1\}$ and $\beta(i)=x_i$. In this case the relation ϱ_a expresses "direct right adjacency". The corresponding graphical representation of d_w has the following form, if we represent nodes by circles with their denotation outside and their label inside the circle:

Fig. 8

Let $d(\Sigma_V,\Sigma_E)$ denote the set of all l-graphs over the alphabets Σ_V, Σ_E and d_ε the empty graph, i.e. the graph with empty set of nodes. As we are not interested to distinguish between two graphs which have the same structure but different node denotations we define:

Def. 3.2: Graphs $d\in d(\Sigma_V,\Sigma_E)$ and $d'\in d(\Sigma_V,\Sigma_E)$ are called equivalent (abbr. $d\equiv d'$ or $d\underset{f}{\equiv}d'$) iff there is a bijective mapping f: $K\rightarrow K'$ with

- . $(k_1,k_2)\in\varrho_a \Leftrightarrow (f(k_1),f(k_2))\in\varrho_a'$, for any $a\in\Sigma_E$,
- . $\beta=\beta'\circ f$. 3)

The relation $\equiv$ is an equivalence relation over any subset of $d(\Sigma_V,\Sigma_E)$. Let D denote the class of all graphs of $d(\Sigma_V,\Sigma_E)$ equivalent to d, and analogously for other names of graphs. So, $d\in D$, $d_1\in D_1$, $d'\in D'$ etc. The graphical representation of a class of l-graphs is the same as the graphical representation of a representative but without denotations for nodes. Let be $D(\Sigma_V,\Sigma_E):=d(\Sigma_V,\Sigma_E)/\equiv$.

2) In [22]-[26] we have taken natural numbers as node labels, which is no restriction. The resulting graphs have been called n-diagrams or n-graphs.

Def. 3.3: The graph $d' \in d(\Sigma_V,\Sigma_E)$ is a <u>partial graph</u> of $d \in d(\Sigma_V,\Sigma_E)$ (abbr. $d' \sqsubseteq d$) iff $K' \subseteq K$, $\varrho'_a \subseteq \varrho_a$, for any $a \in \Sigma_E$, $\beta' = \beta|_{K'}$. [4]

The graph $d' \in d(\Sigma_V,\Sigma_E)$ is a <u>subgraph</u> of $d \in d(\Sigma_V,\Sigma_E)$ (abbr. $d' \subseteq d$) iff $d' \sqsubseteq d$ and $\varrho'_a = \varrho_a \cap (K' \times K')$, for any $a \in \Sigma_E$, i.e. all edges connecting nodes of K' belong to d'.

Let $d', d'' \in d(\Sigma_V,\Sigma_E)$ and $\beta'(k) = \beta''(k)$ for all $k \in K' \cap K''$. The 1-graph $d = (K' \cup K'', (\varrho_a)_{a \in \Sigma_E}, \beta)$ with $\varrho_a = \varrho'_a \cup \varrho''_a$ for any $a \in \Sigma_E$, $\beta(k) = \beta'(k)$ for $k \in K'$, and $\beta''(k)$ for $k \in K''$ is called the <u>union graph</u> of d' and d'' and shortly written as $d = d' \cup d''$.

Let $d' = (K', (\varrho'_a)_{a \in \Sigma_E}, \beta')$ and $K'' \subseteq K'$. Then the subgraph of d' with the node set K'' is called the subgraph <u>generated by K''</u> and written as $d'(K'')$.

As already mentioned in section 1 and elucidated in section 2, a graph production is a triple $p = (d_l, d_r, C)$ with d_l and d_r being the mother node and daughter graph of p respectively, and C an algorithmic specification which determines how to connect daughter graphs depending on the connection between mother nodes. In our approach, C is constructed in the following way: For incoming edges of a certain label, C specifies which nodes of the daughter graph are target nodes for these edges. Moreover, C has to determine those daughter graphs where these edges are to come from. This determination is done in the state before the derivation is applied, i.e. the daughter graphs are distinguished by specifying the corresponding mother nodes. In the same way the connection transformation C indicates also the source nodes within a daughter graph for outgoing edges of a certain label and, furthermore, those daughter graphs where these edges are to point to. Thus, the connection transformation can be considered as specifying half-edges, and an edge between two daughter graphs is inserted iff two half-edges of the same label fit together.

To describe this specification of half-edges very general, we define special strings which we call operators. These operators are taken over an alphabet Σ consisting of the node label alphabet Σ_V , basic operators L_a and R_a for each symbol of the underlying edge label alphabet Σ_E, and, furthermore, we have some operations allowing us to create new operators from given ones. Operators are a tool to specify subsets of the node set of a given 1-graph in a very general way.

Def. 3.4: Strings over Σ with $\Sigma := \Sigma_V \cup \{L_a | a \in \Sigma_E\} \cup \{R_a | a \in \Sigma_E\} \cup \{C, \cup, \cap,), (\}$ are called <u>operators</u> iff they are formed according to the following rules:

- L_a and R_a are operators, for any $a \in \Sigma_E$.
- If A is an operator, then (CA) and $(\bar{v}A)$ are operators for any $v \in \Sigma_V^+$. [5]
- If A and B are operators, then also AB, $(A \cup B)$, and $(A \cap B)$.

Let $op(\Sigma)$ denote the set of all strings of Σ generated by this recursive scheme.

Def. 3.5: Let $d \in d(\Sigma_V,\Sigma_E)$, d' with node k' a one-node subgraph of d , $k \in K$ an arbitrary node and $A \in op(\Sigma)$. An <u>interpretation</u> $A^{d',d}(k)$ of A yields a subset of the node set K. [6] The definition of an interpretation of an operator is recursive accor-

---

3) Composition of mappings is read from right to left.

4) $\beta|_{K'}$ is the restriction of $\beta: K \to \Sigma_V$ to $K' \subseteq K$.

5) Σ_V^+ is the set of all strings with symbols of Σ_V which have at least length 1.

6) For the definition of an interpretation there is no need to assume that d' is a one-node graph, but in the context of L-systems we are only interested in replacing one-node subgraphs.

ding to Def. 3.4:

$L_a^{d',d}(k) := \{\underline{k} \mid \underline{k} \in K - \{k'\} \wedge (\underline{k}, k) \in \rho_a\}$

$R_a^{d',d}(k) := \{\underline{k} \mid \underline{k} \in K - \{k'\} \wedge (k, \underline{k}) \in \rho_a\}$

$L_a^{d',d}(k)$ specifies the set of source nodes $\neq k'$ of a-labelled edges which end in k and $R_a^{d',d}(k)$ the set of target nodes $\neq k'$ of a-labelled edges leaving k.

$(\complement A)^{d',d}(k) := \{\underline{k} \mid \underline{k} \neq k' \wedge \underline{k} \in K - A^{d',d}(k)\}$

determines the complement of $A^{d',d}(k)$ and

$(v_1 \dots v_m A)^{d',d}(k) := \{\underline{k} \mid \underline{k} \in A^{d',d}(k) \wedge \beta(\underline{k}) \in \{v_1, v_2, \dots, v_m\}\}$

yields the subset of $A^{d',d}(k)$ the nodes of which are labelled with one of the symbols of the string $v_1 \dots v_m$.

$AB^{d',d}(k) := \{\underline{k} \mid \exists \underline{\underline{k}} (\underline{\underline{k}} \in B^{d',d}(k) \wedge \underline{k} \in A^{d',d}(\underline{\underline{k}}))\}$

$(A \cap B)^{d',d}(k) := A^{d',d}(k) \cap B^{d',d}(k)$

$(A \cup B)^{d',d}(k) := A^{d',d}(k) \cup B^{d',d}(k)$

specify the set of nodes $\neq k'$ of K which is generated by sequencing, parallel connection, or branching of operators.

It is easy to show that for any quadruple A,d',d,k,satisfying the above conditions,the interpretation $A^{d',d}(k)$ is uniquely determined. The following examples may familiarize the reader with the notion of operators:

$(\complement R_i)^{d',d}(k)$ is the set of all nodes $\neq k'$ which are not target nodes of i-labelled edges originating in k.

$L_b R_j^{d',d}(k)$ is the set of all nodes $\underline{k} \neq k'$ connected with k by a chain of the form of Fig. 9, where $\underline{\underline{k}} \neq k'$

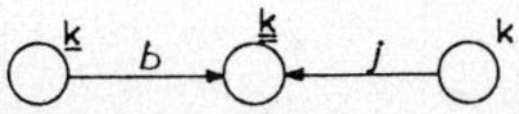

Fig. 9

$((v_1 L_i) \cup (v_2 R_j))^{d',d}(k)$ is the set of all nodes $\neq k'$, which are labelled with $v_1 \in \Sigma_V$ and which are source nodes of i-labelled edges ending in k, or which are labelled with $v_2 \in \Sigma_V$ and which are target nodes of j-labelled edges originating in k.

$(v_1 (L_i \cap L_j))^{d',d}(k)$ is the set of all nodes $\neq k'$ labelled with $v_1 \in \Sigma_V$ which are source nodes of both i-labelled and j-labelled edges which terminate in node k.

Remark 3.6: The set $A^{d',d}(k)$ can be <u>empty</u> for any interpretation, e.g. the operator $O = (v_1 (v_2 L_a))$ yields the empty set for any $v_1 \neq v_2$, $v_1, v_2 \in \Sigma_V$ because of the uniqueness of node labels. On the other hand, there are operators which, for any interpretation, yield <u>all</u> nodes of the graph except k', e.g. the operator $I = (\complement O)$. One can define two operators to be equivalent iff they yield the same set of nodes under any interpretation. Then a lot of rules hold true which the reader can find in [22].
As for iterated unions or iterated intersections the resulting operators are always equivalent, we omit the corresponding brackets, i.e. we write $(A \cup B \cup C)$ for $((A \cup B) \cup C)$ and $(A \cup (B \cup C))$. Furthermore we omit the outermost pair of brackets if the operator is not a part of another operator, i.e. we write vA, $\complement A$, $A \cup B \cup C$, for (vA), $(\complement A)$, $(A \cup B \cup C)$ but of course not $A \cap B)(C \cap D$ for $(A \cap B)(C \cap D)$.
More complicated operators are $Min := \complement(R_V) I$ and $Max := \complement(L_V) I$ with

$R_V := \bigcup_{a \in \Sigma_E} R_a$, $L_V := \bigcup_{a \in \Sigma_E} L_a$ which yield all maximal and all minimal nodes respectively of a host graph besides the single node k' which is to be specified. Trivially, the operator $Is := Max \cap Min$ then determines all isolated nodes.

Def. 3.7: A graph production is a triple $p=(d_l,d_r,C)$ where $d_r\in d(\Sigma_V,\Sigma_E)$ is an arbitrary graph (possibly empty), $d_l\in d(\Sigma_V,\Sigma_E)$ is a one-node graph, both d_l and d_r are loop free [7], and a connection transformation $C=((l_a,r_a))_{a\in\Sigma_E}$ the components of which (called connection components) have the following form:

$$l_a=\bigcup_{\lambda=1}^{q} A_\lambda(k_l)\times\{k_\lambda\},\qquad r_a=\bigcup_{\lambda=1}^{p}\{k_\lambda\}\times A_\lambda(k_l),\qquad a\in\Sigma_E,\ p,q\geq 1,$$

Where k_l is the node of d_l, k_λ is an arbitrary node of K_r, and A_λ is an operator. The graphs d_l and d_r are called mother node and daughter graph of p respectively.

Example 3.8: An example of a graph production is given by [8]:

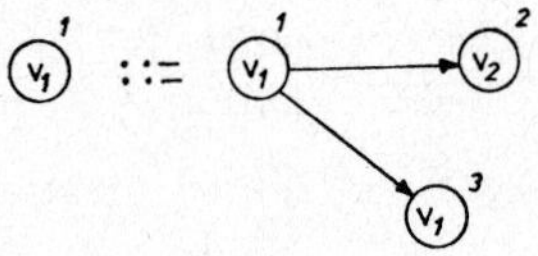

C: $l=(R\cup L)(1)\times\{1\}\cup(R\cup L)(1)\times\{2\}$

$r=\{3\}\times(\complement R)(1)\cup\{3\}\times(L\cap R)(1)$

Fig. 1o

Up till now the connection components of productions are not interpreted, as all corresponding operators are not interpreted, i.e. the host graph and the mother node are not specified. Thus, the connection components are only well-formed strings. The mechanism of establishing connecting edges between daughter graphs can be completely understood only after reading Def. 3.15.

To simplify the notation of connection components we introduce the following abbreviations: [9]

Abbreviation 3.9:

$(A(k_l);k_\lambda):=A(k_l)\times\{k_\lambda\}$, $\qquad (k_\lambda;A(k_l)):=\{k_\lambda\}\times A(k_l)$

$(A(k_l);k_1,k_2,\dots,k_n):=\bigcup_{\lambda=1}^{n}(A(k_l);k_\lambda)$, $\qquad (k_1,k_2,\dots,k_n;A(k_l)):=\bigcup_{\lambda=1}^{n}(k_\lambda;A(k_l))$

Def. 3.1o: Two productions $p=(d_l,d_r,C)$ and $p'=(d'_l,d'_r,C')$ are called equivalent (abbr. $p\equiv p'$) iff $d_l\equiv_f d'_l$, $d_r\equiv_g d'_r$, and C' is the result of C after the following substitution: k_l is replaced by $f(k_l)$ and all node denotations of K_r are replaced by their image in K'_r according to g.

Example 3.11: Trivially the following production is equivalent to the production of Example 3.8 (we use Abbreviation 3.9 and Remark 3.6).

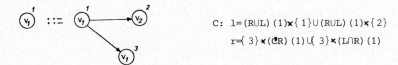

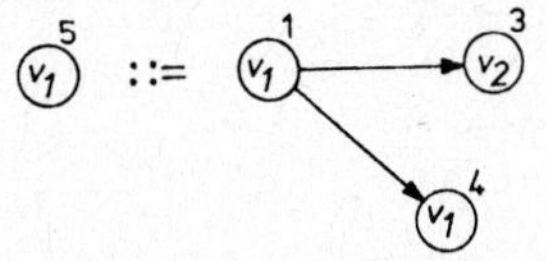

C: $l=(R\cup L(5);1,3)$

$r=(4;\complement R(5))\cup(4;L\cap R(5))$

Fig. 11

---

7) The condition of loop freeness is not a restriction, as reflexive parts of the relations can be expressed by splitting the corresponding node labels. On the other hand this condition ensures a graph L-scheme (cf. Def. 3.14) not to have a lot of unnecessary productions.

8) In this example we assume an edge label alphabet of only one symbol, which is therefore omitted.

Def. 3.12: Let P a set of productions according to Def. 3.7. A production p <u>is contained in P up to equivalence</u> (abbr. $p \overset{\in}{\equiv} P$) iff there is a $p' \in P$ with $p \equiv p'$.

Def. 3.13: A set of productions $\{p_1,\ldots,p_q\}$ (which may contain equivalent productions) is called <u>node disjoint</u> iff the node denotations of all mother nodes are different and the node sets of all daughter graphs are disjoint.

Def. 3.14: A <u>graph L-scheme</u> (shortly GL-scheme) is a triple $S=(\Sigma_V,\Sigma_E,P)$ where

- Σ_V,Σ_E are finite sets, the sets of <u>node</u> and <u>edge labels</u>,
- P is a finite set of graph productions according to Def. 3.7 which is <u>complete</u> in the following sense: for any node label $b \in \Sigma_V$ there is a production in P with b as label of its mother node.

If for any $b \in \Sigma_V$ there is exactly one production with b as mother node label, then is called <u>deterministic</u>, and if the daughter graphs of all productions are different from d_ε then the scheme is called <u>propagating</u>.

Def. 3.15: Let $S=(\Sigma_V,\Sigma_E,P)$ be a graph L-scheme and $d \in d(\Sigma_V,\Sigma_E)-\{d_\varepsilon\}$, $d' \in d(\Sigma_V,\Sigma_E)$. The graph d' is <u>directly derivable</u> from d (abbr. $d \xrightarrow[S]{} d'$) iff there is a set of node disjoint productions $\{p_1,\ldots,p_k\}$ with $p_\mu \overset{\in}{\equiv} P$ for $1 \leq \mu \leq k$ such that

a) $K=\bigcup_{\mu=1}^{k} K_{l\mu}$, $d_{l\mu} \subseteq d$

b) $K'=\bigcup_{\mu=1}^{k} K_{r\mu}$, $d_{r\mu} \subseteq d'$

c) The connection between two daughter graphs is as follows:
An a-labelled edge is going from $k_1 \in K_r$ to $k_2 \in K'_r$ iff the following two conditions hold
c1) the connection component l_a of production p' contains a part $(A(k'_l);k_2)$ where $k_l \in A^{d'_l,d}(k'_l)$
c2) the connection component r_a of the production p contains a part $(k_1;B(k_l))$ where $k'_l \in B^{d_l,d}(k_l)$.

Remark 3.16: Condition a) means that for all one-node subgraphs of d we have a corresponding production of P up to equivalence which has this subgraph as mother node.
Condition b) ensures that all daughter graphs of the applied productions are subgraphs of d' and that their node sets are a partition of the node set K'.
Condition c) may be elucidated by the following figure:

Fig. 12.a

mother nodes

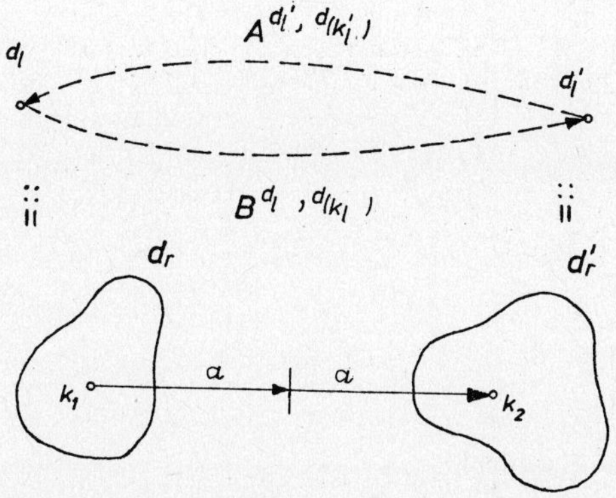

Fig. 12.b

daughter graphs

---

9) In this special case where productions only have one-node left hand sides the denotation of mother nodes is unnecessary. Nevertheless, we write the connection transformation in this more complicated manner to have a notation which is applicable for both sequential and parallel rewriting (and in the latter case for rewriting of more complex subgraphs than single nodes as indicated in section 1).

Each of the above two conditions c1 and c2 can be regarded as generating a "half-edge": The first condition specifies that an edge with label a starting anywhere in d_π is to go into node k_2 of d'_π. Thus the right half-edge between d_π and d'_π is determined. While the node k_2 is explicitly determined, the determination of the daughter graph d_π as source is done in the graph before the derivation step is applied by yielding the corresponding mother node d_l by the first part of the connection part (cf. Fig. 12.a). Analogously condition c2 determines the left half-edge by specifying that an edge with label a, originating in k_1 of d_π is to go into a node of d'_π. An a-labelled edge is connecting k_1 with k_2 iff both half-edges have the same label and fit together as indicated in Fig. 12.b. Furthermore, condition c shows that connection parts whose operator is the empty operator O (cf. Remark 3.6) can be omitted, because they cannot yield any mother node and consequently they cannot specify any daughter graph being source or target of an edge. In the same way we can omit connection components which contain only empty operators.

Derivation can be easily generalized to classes of l-graphs by the following definitions:

Def. 3. 17: Let S be a GL-scheme. $D' \in D(\Sigma_V, \Sigma_E)$ is <u>directly derivable</u> from $D \in D(\Sigma_V, \Sigma_E)$ iff there exists a $d \in D$ and a $d' \in D'$ such that $d \underset{S}{\rightarrow} d'$.

Def. 3. 18: Let S be a GL-scheme and $\overset{*}{\underset{S}{\rightarrow}}$ be the reflexive and transitive closure of $\underset{S}{\rightarrow}$ on $d(\Sigma_V, \Sigma_E)$ and $D(\Sigma_V, \Sigma_E)$ respectively. A graph d' is called <u>derivable</u> from d (abbr. $d \overset{*}{\underset{S}{\rightarrow}} d'$) iff $d \overset{*}{\underset{S}{\rightarrow}} d'$. D' is derivable from D iff there exists a $d \in D$ and a $d' \in D'$ with $d \overset{*}{\underset{S}{\rightarrow}} d'$.

The following three lemmas are trivial consequences of the definition of direct derivation.

Lemma 3. 19: Let be $D \underset{S}{\rightarrow} D'$ with S as above, and let $d \in D$ and $d' \in D'$. This implies $d \underset{S}{\rightarrow} d'$.

Lemma 3. 2o: Let S be a GL-scheme and $d \in d(\Sigma_V, \Sigma_E) - \{d_\varepsilon\}$. Then there exists a graph $d' \in d(\Sigma_V, \Sigma_E)$ with $d \underset{S}{\rightarrow} d'$.

In an arbitrary GL-scheme S there need not be only one successor of a class of l-graphs under the relation $\underset{S}{\rightarrow}$. This is due to the fact that there may be different daughter graphs which can rewrite one mother node, or that there may be different connection transformations for a given pair of mother node and daughter graph. However, for deterministic GL-schemes we trivially have:

Lemma 3. 21: Let S be a deterministic GL-scheme. If $d \rightarrow d'$ then D' is uniquely determined.

Def. 3. 22: A <u>graph L-system</u> (shortly GL-system) is a pair $G=(S, d_\omega)$ where $S=(\Sigma_V, \Sigma_E, P)$ is a GL-scheme and $d_\omega \in d(\Sigma_V, \Sigma_E) - \{d_\varepsilon\}$. The graph d_ω is called <u>axiom graph</u> and has no loops. [7] Derivation is defined as derivation in the underlying scheme S and is denoted as $d \underset{G}{\rightarrow} d'$, $D \underset{G}{\rightarrow} D'$, $d \overset{*}{\underset{G}{\rightarrow}} d'$, $D \overset{*}{\underset{G}{\rightarrow}} D'$, and by the same notation without G if no confusion can occur.

Def. 3. 23: The <u>language</u> of a GL-system is defined as
$\mathcal{L}(G) := \{D \mid D \in D(\Sigma_V, \Sigma_E) \text{ and } D_\omega \overset{*}{\underset{G}{\rightarrow}} D\}$.
Furthermore, a set of graph classes is called a <u>GL-language</u> iff there exists a GL-system generating it.

For the introduction of extended graph L-systems we can split the underlying node label alphabet Σ_V into a terminal and a nonterminal one. The interpretation of nonterminal structures in biological applications is, that the description of development in time interval (n,n+1) cannot be accomplished by one derivation step. So, we need a sequence of intermediate graphs which have no significance in nature but are necessary to get the next terminal graph which represents the organism in state n+1. Following this idea it seems to be natural to split not only the alphabet of node labels but also the alphabet of edge labels Σ_E into a terminal and nonterminal one in order to increase the generative and manipulative power of the system.

Def. 3. 24: An extended graph L-system (shortly EGL-system) is a triplet $H=(G,\Delta_V,\Delta_E)$ where $G=((\Sigma_V,\Sigma_E,P),d_\omega)$ is a GL-system, $\Delta_V \subseteq \Sigma_V$ and $\Delta_E \subseteq \Sigma_E$. The subset Δ_V is called the terminal node label alphabet, $\Sigma_V - \Delta_V$ the nonterminal node label alphabet, Δ_E the terminal edge label alphabet, and $\Sigma_E - \Delta_E$ the nonterminal edge label alphabet respectively. Derivation is defined as derivation in the underlying GL-system and is denoted as $d \underset{H}{\rightarrow} d'$, $d \overset{*}{\underset{H}{\rightarrow}} d'$, etc. (Thus an arbitrary graph derived from d_ω by H can have terminal and nonterminal nodes and edges.)
The language of a EGL-system is defined as
$\mathcal{L}(H) := \{ D \mid D \in D(\Delta_V, \Delta_E) \text{ and } D_\omega \overset{*}{\underset{H}{\rightarrow}} D \}$.

So, classes of graphs contained in the language of an EGL-system only have terminal node labels and terminal edge labels.

Finally, we can generalize the notion of TOL-systems to get table graph L-systems which, for biological applications, correspond to changes of development within the growth of an organism.

Def. 3. 25: A table graph L-system (shortly TGL-system) is a tuple $G=(\Sigma_V,\Sigma_E,\mathcal{P},d_\omega)$ where $\mathcal{P}=\{P_1,\dots,P_m\}$, and $S_i=(\Sigma_V,\Sigma_E,P_i)$ are GL-schemes for $1 \leq i \leq m$, called the component schemes of G. The production sets P_i are called tables. Direct derivation is defined as direct derivation in one of the component schemes, i.e. all applied productions are chosen from only one table. The language of a TGL-system is defined as in Def. 3. 23.

It is easy to see how the definitions of EGL-systems and TGL-systems must be combined to get extended table graph L-systems if the necessity for this definition arises.

The generative power of the above parallel rewriting mechanism may be emphasized by the following examples, where we indicate only those connection components, which generate the edges between the daughter graphs of the derived graph given below.

Examples 3.26 Besides the trivial capability to delete connections, i.e. two mother nodes are connected by an edge while the corresponding two daughter graphs are not connected, the above parallel mechanism can perform the following manipulations in one derivation step:

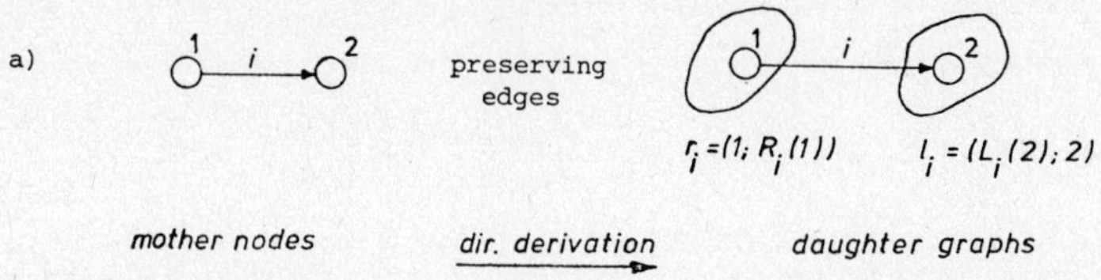

Fig. 13

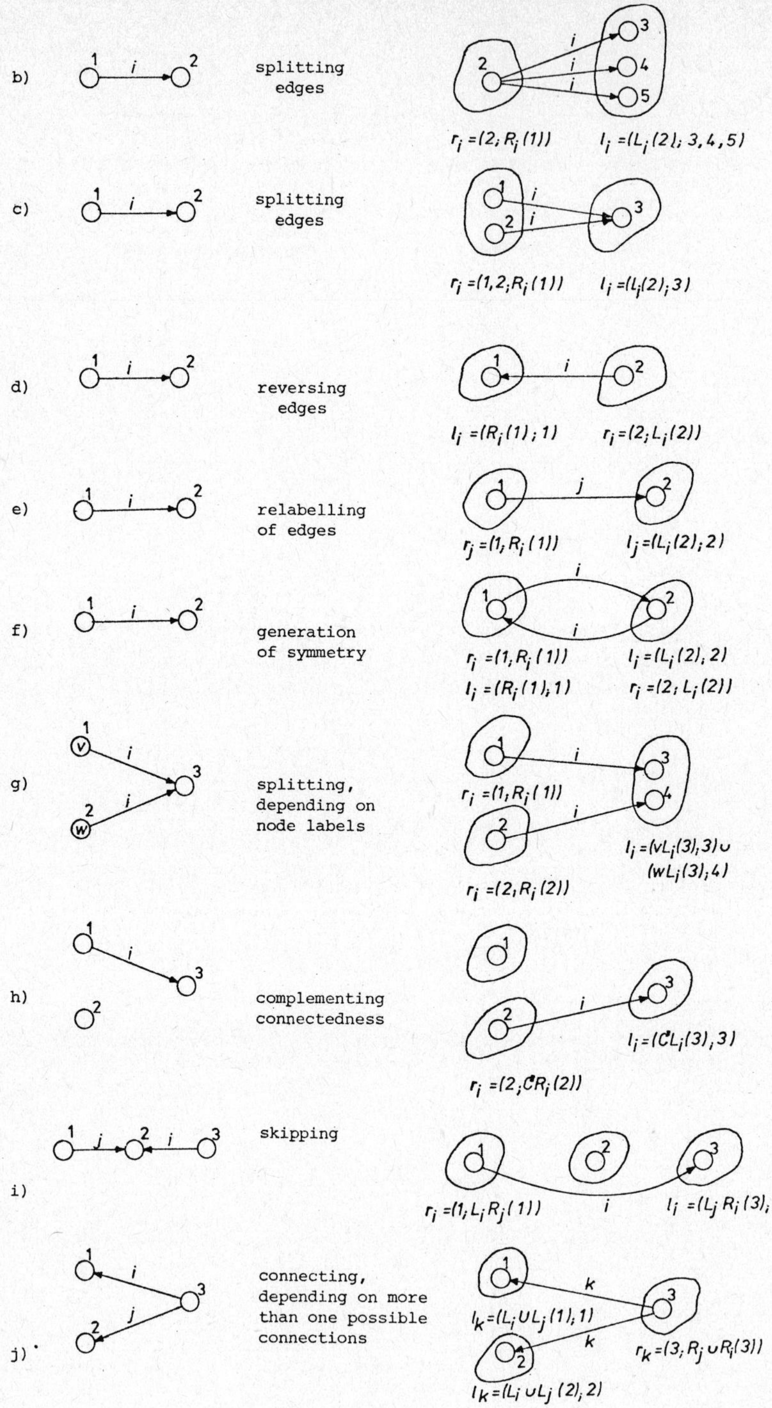
b)
splitting edges
r_i =(2; R_i (1))
l_i =(L_i (2); 3,4,5)
c)
splitting edges
r_i =(1,2; R_i (1))
l_i =(L_i(2); 3)
d)
reversing edges
l_i =(R_i (1); 1)
r_i =(2; L_i (2))
e)
relabelling of edges
r_j =(1; R_i (1))
l_j =(L_i (2); 2)
f)
generation of symmetry
r_i =(1; R_i (1))
l_i =(L_i (2); 2)
l_i =(R_i (1); 1)
r_i =(2; L_i (2))
g)
splitting, depending on node labels
r_i =(1; R_i (1))
r_i =(2; R_i (2))
l_i =(vL_i(3); 3) ∪ (wL_i(3); 4)
h)
complementing connectedness
l_i =(CL_i(3); 3)
r_i =(2; CR_i (2))
i)
skipping
r_i =(1; L_j R_j (1))
l_i =(L_j R_i (3); 3)
j)
connecting, depending on more than one possible connections
l_k =(L_i ∪ L_j (1); 1)
r_k =(3; R_j ∪ R_i(3))
l_k =(L_i ∪ L_j (2); 2)

Fig. 13

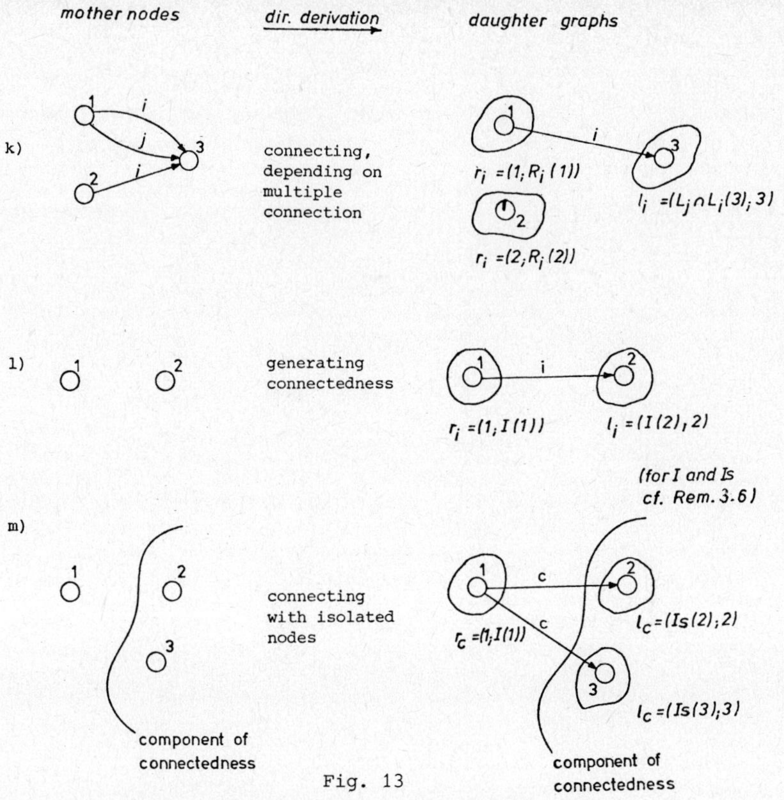

Fig. 13

4. SOME RESULTS ON GRAPH REWRITING SYSTEMS

In the following we only quote some of the results known for graph rewriting systems and comment them. Proofs are not presented here, the reader is asked to refer to [22]-[26].

For parallel rewriting on graphs, as introduced in section 3, we get the following diagram of inclusions (cf. [25]):

(1)
$$\begin{array}{ccccccc} [GL] & \subset & [EGL] & \supset & [PEGL] & \supset & [PTGL] \\ & & \cup & \cup & \supset & & \\ & & [IL] & [TOL] & [EOL] & & \end{array}$$

where [GL],[EGL],[PEGL], and[PTGL] represent the class of GL-languages, EGL-languages, propagating EGL-languages, and propagating TGL-languages, respectively, and where [IL],[TOL],[EOL] denotes the class of corresponding string L-languages interpreted as languages of linearly ordered graphs as indicated in Fig. 8.

Most of the inclusions in diagram (1), e.g. [GL]⊂[EGL],[PEGL]⊂[EGL], [EOL]⊂[PEGL], and [TOL]⊂[PEGL], can easily be proved. The two remaining results [IL]⊂[EGL],[PTGL]⊂[PEGL]are more difficult to infer. The first one shows that the derivation mechanism of string L-systems with interaction (all symbols of a string are rewritten simultaneously iff each symbol has a left and right context of a certain depth as specified in the corresponding string production) can be simulated by context free graph rewriting, i.e. the context is not a part of the left and right hand side of the graph production. This result is due to the fact that the operator concept, contained in the notation of connection transformations, allows us to check the left and right context whether it has the form which is specified in the corresponding string IL-production. The second result [PTGL] $\subset$ [PEGL] means that programming of graph development by using the table rewriting mechanism, can be simulated by the usual mechanism if one allows nonterminal nodes and edges. For these two results, as well as for the result which are mentioned below, the programming capability, enabled by the operator concept of the connection components (or embedding components in the sequential case) is highly important. Furthermore, by simulating e.g. table mechanism or string IL-mechanism by extended rewriting, this simulation is done in a sequence of steps. To avoid that, in the simulating system, productions are applied in an order not consistent with the simulation, the introduction of special nonterminal edges has been very fruitful. These edges, called "false" edges, are generated if a "wrong" production was applied, and never occur, if a "correct" simulation has taken place. The connection components for these edges are built in such a way that false edges can never be deleted if they have been generated once. Thus, from a false graph, a graph of the language of an EGL-system can never be derived.

As already announced in the introduction, productions, as defined in 3.7, can be used for sequential rewriting on graphs, too. The only difference is that in the graph grammar case left hand sides contain more than one node. The third component of the graph productions, called embedding transformation in this case, can be interpreted in the following way: In the components l_a, $a \in \Sigma_E$ which specify incoming edges, nodes of the right hand side are explicitly given. These nodes are target nodes of incoming edges of label a. The source nodes, which lie in the unchanged part of the host graph (cf. Fig. 1), are specified by an operator expression. Thus, in sequential rewriting, embedding components determine edges, not half-edges as in the parallel case. Analogously, for components r_a, $a \in \Sigma_E$, the source nodes for outgoing edges are explicitly given while the target nodes are determined by an operator expression.

In common string grammars, a hierarchy of language classes exists according to a successive restriction of the form a production may have. One can generalize these notions to labelled graphs [1o]. This yields a hierarchy of classes of sequential graph languages (cf [22],[23]):

(2) $\quad [U] \supset [M] = [CS] \supseteq [CF] = [N] \supseteq [R] = [RN]$

where [U],[M],[CS],[CF],[N],[R], and [RN] denotes the class of languages of graph grammars which is unrestricted, monotone, context sensitive, context free, context free in normal form, regular, and regular in normal form, respectively. In the same way as in the parallel case there are rather evident parts of that hierarchy as $[U] \supset [M[$, $[M] \supseteq [CS]$, $[CS] \supseteq [CF]$, $[CF] \supseteq [N]$, $[CF] \supseteq [R]$, and $[R] \supseteq [RN]$. The remaining results $[M] \subseteq [CS]$, $[CF] \subseteq [N]$, and $[R] \subseteq [RN]$ are normal form theorems for monotone, context free, and regular graph grammars respectively. In each of these cases, a production is splitted into a sequence of productions of a simpler form, such that this sequence has the same effect as the production it originates from. To avoid that, in this simulation, productions are applied in another order than given in this sequence, again nonterminal edges are used. So, in the same way as in the parallel case, these theorems are based on the concept of operators in the embedding transformation and the idea of false edges. It is clear that a simplification of the form of left and right hand sides on the other hand complicates the grammar: We need additional node and edge labels and sometimes more complicated embedding transformations. This argument, however, is also true for all normal form theorems known in the theory of string languages.

The possibility to program graph development by these two concepts can be used to prove a result which has no analogon in the theory of string languages, and which interrelates the two language hierarchies (1) and (2) (cf. [26]):

1o) There are more than one possible generalizations for each definition, especially for the notion of context sensitivity and regularity on graphs.

(3) [PEGL] = [CF].

That we get [PEGL]=[CF] and not [EGL]=[CF] is due to the fact that context freeness was defined in [22],[23] without deletion. The equation [PEGL]=[CF] means that a context free sequential derivation step can be simulated by two parallel derivation steps and, vice versa, that a parallel derivation step can be simulated by a sequence of sequential steps.

5. IMPLICIT VERSUS EXPLICIT PARALLEL PROGRAMMING

In section 3 we have introduced a parallel rewriting mechanism by which connections between daughter graphs are generated implicitly: The connection transformation is a part of each production and, because an edge between two daughter graphs relates to both daughter graphs, the connection transformation cannot generate complete edges. In section 3 we have elucidated this fact with the aid of an half-edge mechanism. We have argued that each connection transformation generates half-edges of different labels, and edges are built up between two daughter graphs iff two half-edges of the same label fit together.

In [6] an approach of parallel rewriting on graphs is given which we would like to call explicit parallel programming. A simplified and modified form which forgoes the "matching hands" and "open hands" of [6] is presented in the sequel. The idea of this approach is the following:
We have two kinds of graph productions, replacement rules and connection rules. Replacement rules consist of pairs of mother node and daughter graph. The connection rules have an edge with source and target node as left hand side, called mother edge, and an arbitrary graph as right hand side. The node set of this graph is divided into two disjoint parts such that the right hand side consists of two disjoint subgraphs (which are generated by these node sets), the "source graph" and the "target graph", and edges which connect nodes of these two subgraphs. These connecting edges link the daughter graphs together which are inserted for the two mother nodes of the mother edge if the source graph is embedded in the daughter graph corresponding to the source node of the mother edge, and if the target graph is embedded in the daughter graph corresponding to the target node of the mother edge.

Def. 5.1: Let Σ_V,Σ_E be as above the node label alphabet and the edge label alphabet. A <u>replacement rule</u> is a pair $r=(d_l,d_r)$ with $d_l \in d(\Sigma_V,\emptyset)$ being a single labelled node and $d_r \in d(\Sigma_V,\Sigma_E)$ being an arbitrary loopless graph. As above, d_l is called <u>mother node</u>, d_r <u>daughter graph</u>.
A <u>connection rule</u> is a pair $c=(d_e,d_s)$ with $d_e \in d(\Sigma_V,\Sigma_E)$ called <u>mother edge</u> being a graph consisting of two nodes k_1 and k_2 which are labelled with a symbol of Σ_V, and an edge from k_1 to k_2 with label from Σ_E. In the graph $d_s \in d(\Sigma_V,\Sigma_E)$ which is called <u>stencil</u> the node set K_s consists ot two disjoint subsets K_{k_1} and K_{k_2}. The subgraphs $d_s(K_{k_1})$ and $d_s(K_{k_2})$ of d_s are called <u>source graph</u> and <u>target graph</u> respectively. 11)
Two replacement rules $r=(d_l,d_r)$ and $r'=(d_l',d_r')$ are called <u>equivalent</u> iff $d_l \equiv d_l'$, and $d_r \equiv d_r'$. In the same way equivalence of connection rules is defined. A set of replacement rules or connection rules is called <u>node disjoint</u> iff the node sets of all left hand sides are disjoint and the node sets of all right hand sides are disjoint.

---

11) The stencil and the axiom graph of Def. 5.2 are loop free too.

Def. 5.2: A graph L-system with explicit programming is defined as $G=(\Sigma_V,\Sigma_E,RR,CR,d_\omega)$

- where Σ_V,Σ_E and d_ω is as above the alphabet of node and edge labels, and the axiom graph over Σ_V,Σ_E,
- RR is a finite set of replacement rules as defined above, which is complete in the following sense that for each $a\in\Sigma_V$ there is a replacement rule with a as mother node label,
- CR is a finite set of connection rules as defined above, which is complete in the sense that for each triple (v_1,a,v_2) $v_1,v_2\in\Sigma_V$, $a\in\Sigma_E$ there is a connection rule with v_1 and v_2 being the labels of the source and target node, and a being the label of the connecting edge.

Def. 5.3: Let G be a graph L-system with explicit programming. The l-graph $d'\in d(\Sigma_V,\Sigma_E)$ is called explicitly direct derivable from $d\in d(\Sigma_V,\Sigma_E)$ in G (abbr. $d \underset{G}{\rightarrow} d'$) iff there is a set of replacement rules $\{r_1,\dots,r_k\}$ and a set of connection rules $\{c_1,\dots,c_q\}$ such that

- r_μ $1\leq\mu\leq k$ is equivalent to a $r\in RR$,
- c_λ $1\leq\lambda\leq q$ is equivalent to a $c\in CR$,
- $\{r_1,\dots,r_k\}$ and $\{c_1,\dots,c_q\}$ are node disjoint,
- $d_{l\mu}\subseteq d$, $\biguplus_{\mu=1}^{k} K_{l\mu} = K$, and $d_{r\mu}\subseteq d'$, $\biguplus_{\mu=1}^{k} K_{r\mu} = K'$,
- $\biguplus_{\lambda=1}^{q} d_{e\lambda} = d$, and d' is defined by $\biguplus_{\mu=1}^{k} d_{r\mu} \cup \biguplus_{\lambda=1}^{q} d_{s\lambda}$ with the following restriction: Let d_l and d'_l be two mother nodes of d connected by d_e of $c=(d_e,d_s)$ (mother edge from node k_1 to k_2) then we have $d_s(K_{k_1})\sqsubseteq d_r$ and $d_s(K_{k_2})\sqsubseteq d'_r$.

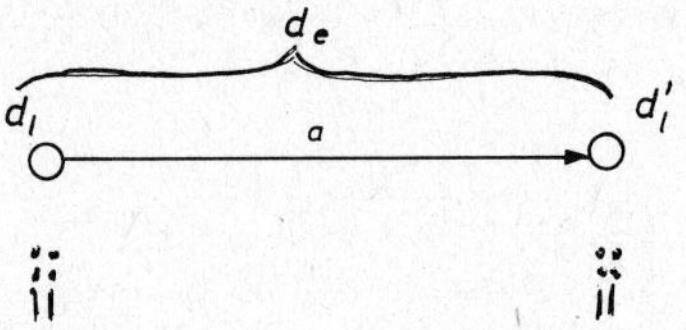

Fig.14.a mother nodes mother edges

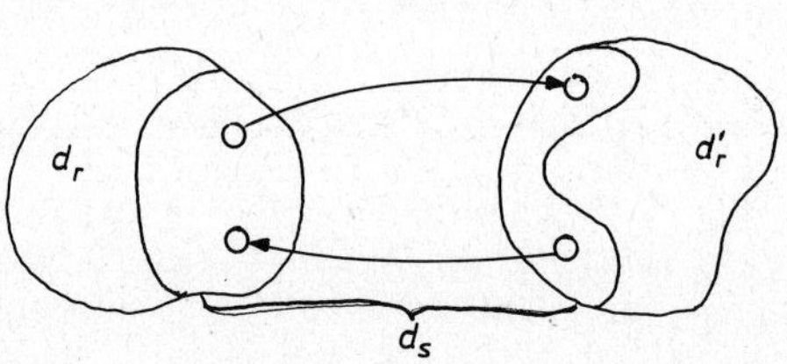

Fig.14.b daughter graphs and stencils

Remark 5.4: Def. 5.4 is trivially extended to classes of graphs from $D(\Sigma_V,\Sigma_E)$. A language of an L-system of Def. 5.3 is defined in the usual way.
It is evident, that the completeness condition of RR and CR is no guarantee that from a graph d always a graph d' can be derived. However, this is the case if $d_s(K_{k_i})\sqsubseteq d_r$, for each daughter graph d_r where the corresponding mother node has the same label as k_i.
In the extreme case that the partial graphs $d_s(K_{k_i})$ are the daughter graphs of k_i then replacement rules are unnecessary as the complete information needed for a derivation step is already given by the connection rules.
In [11] a subcase of the above explicit mechanism is given, where the stencils contain all nodes of the corresponding daughter graph they are embedded in, but no edges between these nodes. The resulting systems are called node substitution parallel graph grammars.
Furthermore, one can define extended systems by splitting the node label alphabet (and edge label alphabet, which is not done in [6]) into a terminal and a nonterminal one, by regarding only those graphs to belong to the language which are terminal.

In [6] a lot of results are given which shall not be presented here, as there was no time to investigate, whether these results are due to the open hands and matching hands of [6] or not, i.e. whether they still hold true for the simplified form of explicit parallel rewriting which is given here.
The bibliography cites most of the approaches on generalized parallel rewriting which are known now. We do not present a comparison between these different definitions, as nearly all of these approaches can be found in [19] and as a careful comparison is too space consuming for this paper.

Remark 5.5: It can be easily understood that the explicit mechanism defined above manipulates only direct neighbourhoods: A connection between daughter graphs can be established iff there is a connection between the corresponding mother nodes. So, if we regard Examples 3.26, manipulations of the form a)-g),j) are possible. The rest of the manipulations of the implicit approach of section 3 is not possible here, i.e. especially the zipping mechanism of Ex. 3.26 i) cannot be programmed in the explicit approach. Concerning the generality of both approaches we state without proof, that the explicit mechanism can be simulated by the extended implicit one. Trivially, the converse is not true.
Besides the question which of the both above mentioned approaches is more general one should remark: Programming graph development explicitly by connection rules is more obvious, especially, if the graphs to be modelled have only few edge labels and if, for a given mother edge, there are only few ways how the corresponding source and target graph can be embedded in the corresponding daughter graph. If both conditions do not hold, then the clearness won by explicit programming, might be destroyed by the great number of connection rules, needed. In the latter case the implicit method which specifies only half-edges might be more advantageous. The reader is encouraged to construct a graph L-system with explicit programming for the development shown in Fig. 7 of section 2. Even in this simple example a lot of connection rules are necessary.

6. FURTHER APPLICATIONS AND IMPLEMENTATION

There are a lot of applications of graph rewriting systems some of which are sketched in the following. For the typical reader of this book these applications might be at the verge of his fields of interest.

The first application of graph grammars was in the field of grammatical pattern recognition which is used to classify fog and bubble chamber pictures. In these applications we have no finite number of picture classes such that any picture can be related to one of these classes. Any point of a picture can be the origin of a subpicture if a disintegration takes place there. On the other hand it is possible to state recursive rules saying how subpictures are generated and how they are connected. Thus, for any class of such pictures, one can give a finite set of rules, i.e. a graph grammar. The advantage of grammatical picture analysis is that in the case of success one has not only a yes-no-decision but a description of the structure of the considered picture. The reader finds a comprehensive bibliography of this topic in [3o].

Apart from graph theoretical theorems special classes of graphs can be characterized by graph grammars which generate them. The reader can find a lot of graph grammars for interesting classes of graphs in [1],[21],[22].

In recent time two studies in the field of incremental compiling have been made [4],[37]. The idea of incremental compiling is that a computer program need not be totally re-compiled if only an increment of this program is changed. To solve this problem, compilation could be made in two steps. In the first step a graph is built up for the program which represents the information a compiler usually keeps in its various lists. In the second step, this graph is translated into a machine program. Changes of the linearly denoted program, by inserting or deleting an increment, induce updating of the graph which can be established by applying graph productions sequentially.

Since there are graphical I/O-devices for computers the question arises whether a two-dimensional notation of algorithms is more adequate to human thinking than a linear one. In [8] a two-dimensional programming language is given,for the definition of which,graph grammars as defined in [35] are used.As it is no problem to state grammars which generate flowcharts,two-dimensional programming can be syntax-directed, i.e. the programmer chooses a rule from a finite set of given ones. In this manner, only syntactically correct flowcharts can be generated and, with a suitable form of the flowchart grammar, this method automatically satisfies the requirements of structured programming.

Complex data and information structures can be regarded as labelled graphs. In [13],[14] graph grammars are used for a precise definition of special types of data structures, which is usually done by the aid of data definition languages. Furthermore, graph rewriting systems may serve as a method to manipulate these data structures which is usually linearly done by data manipulation languages.

Towards an implementation, the first steps have been made. In [3] the implementation of sequential rewriting in the form of [22] is nearly completed. If the subroutine package given there is extended by a subgraph test (which finds the left hand side of a production which is to apply within a host graph) and by a graphical output part, then some of the problems stated above can be practically investigated. The implementation of graph grammars was essentially facilitated because it is based on DATAS[12] with which data can be stored associatively and with which labelled graphs can be easily coded. The implementation of parallel rewriting on graphs as presented in section 3 should be possible without great expenditure, as both rewriting mechanisms use the same definition of a graph production.

7. REFERENCES

[1] Abe,N./Mizumoto,M./Toyoda,J.-I./Tanaka,K.: Web Grammars and Several Graphs, Journ. Comp. Syst. Sci. 7 , 37-65 (1973).

[2] Brayer,J.M./Fu,K.S.: Some Properties of Web Grammars, Techn. Report TR-EE 74-19, Purdue University, Indiana, April 1974.

[3] Brendel,W.: Implementierung von Graph-Grammatiken, Arbeitsber. d. Inst. f. Math. Masch. u. Datenver. 9 , 1, Erlangen, January 1976.

[4] Bunke,H.: Beschreibung eines syntaxgesteuerten inkrementellen Compilers durch Graph-Grammatiken, Arbeitsber. d. Inst. f. Math. Masch. u. Datenver. 7 , 7, Erlangen, December 1974.

[5] Carlyle,J.W./Greibach,S.A./Paz,A.: A two-dimensional generating system modeling growth by binary cell division, Proc. 15th Annual Conf. on Switching and Automata Theory, 1-12 (1974).

[6] Culik,K.II/Lindenmayer,A.: Parallel Rewriting on Graphs and Multidimensional Development, Techn. Report CS-74-22, University of Waterloo, Canada, November 1974.

[7] Culik,K.II: Weighted growth functions of DOL-systems and growth functions of parallel graph rewriting systems, Techn. Report CS-74-24, University of Waterloo, Canada, 1974.

[8] Denert,E./Franck,R./Streng,W.: PLAN2D - Towards a Two-dimensional Programming Language, Lect. Notes in Computer Science 26 , 2o2-213, Berlin: Springer-Verlag 1975.

[9] Ehrig,H./Kreowsky,H.-J.: Parallel Graph Grammars, In: Automata, Languages, Development (ed. by A. Lindenmayer and G. Rozenberg), Amsterdam: North Holland 1976.

[1o] Ehrig,H./Pfender,H./Schneider,H.-J.: Graph-Grammars: An algebraic approach, Proc. 14th Annual Conf. on Switching and Automata Theory, 167-18o (1973).

[11] Ehrig,H./Rozenberg,G.: Some Definitional Suggestions for Parallel Graph Grammars, In: Automata, Languages, Development (ed. by A. Lindenmayer and G. Rozenberg), Amsterdam: North Holland 1976.

[12] Encarnacao,J./Weck,G.: Eine Implementierung von DATAS (Datenstrukturen in Assoziativer Speicherung), Comp. Science Techn. Report A 74-1, University of Saarbrücken, Germany, 1974.

[13] Frühauf,T.: Formale Beschreibung von Informationsstrukturen, Arbeitsber. d. Inst. f. Math. Masch. u. Datenver. 9 , 1, Erlangen, January 1976.

[14] Gotlieb,C.C./Furtado,A.L.: Data Schemata Based on Directed Graphs, Techn. Report 7o, Deptmt. Comp. Science, University of Toronto, Canada, October 1974.

[15] Heibey,H.W.: Ein Modell zur Behandlung mehrdimensionaler Strukturen unter Berücksichtigung der in ihnen definierten Lagerelationen, Report No 15, Inst. f. Informatik, University of Hamburg, Germany, May 1975.

[16] Herman,G.T./Rozenberg,G.: Developmental Systems and Languages, Amsterdam: North Holland 1975.

[17] Lindenmayer,A.: Mathematical Models for Cellular Interactions in Development, Parts I and II, Journ. Theor. Biology 18 , 28o-315 (1968).

[18] Lindenmayer,A.: Developmental Systems without Cellular Interactions, Their Languages and Grammars, Journ. Theor. Biology 3o , 455-484 (1971).

[19] Lindenmayer,A./Rozenberg,G.(Ed.): Languages, Automata, Development, Amsterdam: North Holland 1976.

[2o] Mayoh,B.H.: Multidimensional Lindenmayer Organisms, In: L-Systems (ed. by G. Rozenberg and A. Salomaa), Lect. Notes in Comp. Science 15 , 3o2-326, Berlin: Springer-Verlag 1974.

[21] Montanari,U.: Separable Graphs, Planar Graphs and Web Grammars, Inf. Contr. 16, 243-267 (197o).

[22] Nagl,M.: Formale Sprachen von markierten Graphen, Arbeitsber. d. Inst. f. Math. Masch. u. Datenver. 7 , 4, Erlangen, July 1974.

[23] Nagl,M.: Formal Languages of Labelled Graphs, Computing 16 , 113-137 (1976).

[24] Nagl,M.: Graph Lindenmayer-Systems and Languages, Arbeitsber. d. Inst. f. Math. Masch. u. Datenver. 8 , 1, 16-63, Erlangen, January 1975.

[25] Nagl,M.: On a Generalization of Lindenmayer-Systems to Labelled Graphs, In: Languages, Automata, Development (ed. by A. Lindenmayer and G. Rozenberg), 487-5o8, Amsterdam: North Holland 1976.

[26] Nagl,M.: On the Relation between Graph Grammars and Graph Lindenmayer-Systems, Arbeitsber. d. Inst. f. Math. Masch. u. Datenver. 9 , 1, 3-32, January 1976.

[27] Pavlidis,T.: Linear and Context-free Graph Grammars, Journ. ACM 19 , 11-23 (1972).

[28] Pfaltz,J.L./Rosenfeld,A.: Web Grammars, Proc. Int. Joint Conf. Art. Intell., 6o9-619, Washington 1969.

[29] Rosen,B.: Deriving Graphs from Graphs by Applying a Production, Techn. Report RC 5163, IBM Research Lab Yorktown Heights, December 1974.

[3o] Rosenfeld,A.: Progress in Picture Processing 1969-7o, Computing Surveys 5 , 2, 81-1o8 (1973).

[31] Rosenfeld,A./Milgram,D.L.: Web automata and web grammars, Machine Intelligence 7 , 3o7-324 (1972).

[32] Rozenberg,G./Salomaa,A.(Ed.): L-Systems, Lect. Notes in Computer Science 15 , Berlin: Springer-Verlag 1974.

[33] Rozenberg,G./Salomaa,A.: The Mathematical Theory of L-Systems, Techn. Report DAIMI PB-33, University of Aarhus, Denmark, July 1974.

[34] Salomaa,A.: Formal Languages, New York: Academic Press 1973.

[35] Schneider,H.-J.: Chomsky-Systeme für partielle Ordnungen, Arbeitsber. d. Inst. f. Math. Masch. u. Datenver. 3 , 3, Erlangen, August 197o.

[36] Schneider,H.-J.: A necessary and sufficient condition for Chomsky-productions over partially ordered symbol sets, Lect. Notes in Economics and Math. Systems 78 , 9o-98, Berlin: Springer-Verlag 1973.

[37] Schneider,H.-J.: Syntax-directed Description of Incremental Compilers, Lect. Notes in Computer Science 26, 192-2o1, Berlin: Springer-Verlag 1975.

[38] Schneider,H.-J./Ehrig,H.: Grammars on partial graphs, to appear in Acta Informatica.

D i s c u s s i o n

Concerning the paper of Schürger and Tautu

P.G. Doucet:

The pushing out of one of the daughter cells does solve the problem of division in a fixed grid as long as only one layer of cells is considered. But one is bound to get into trouble if the technique is applied in three dimensions. Moreover, it should be realized quite clearly that this solution may be forced upon us by some choice of formalism rather than by biological considerations, and could therefore be an artefact. The satisfactory description of arrays of dividing automata in two or more dimensions is a notorious problem which has not yet been solved.

Schürger:

In the two- and three-dimensional case we interpret the event of vanishing as a sort of cell lysis (the cytoplasma being destroyed and the necleus being broken into small fragments).

H.E. Wichmann:

My first question concerns your definition of crinkliness given by (3.2). But is it not true that for a circle-like configuration ξ with

"radius" r we should have $C(\xi) = \frac{2\pi r}{4\sqrt{\pi r^2}} = \frac{1}{2}\sqrt{\pi} < 1$?

The second question deals with the disappearance of cells in the plane lattice. If a new cell is built in the inner part of the configuration, an old one will leave its lattice place and, for example, will come into the underlying layer. Did you examine the effect of such a cell flow from top or bottom into the lattice ?

Tautu:

To deal with the first question, we have to show that $C(\xi) > 1$ if $\xi \in \Xi_0$ is not a square array. First observe that $C(\xi) > 1$ if $\xi \in \Xi_0$ is a rectangular array being not a square (all arrays are assumed to be parallel to the coordinate axes). Call $\xi \in \Xi_0$ connected if for all $x, y \in \mathrm{supp}(\xi)$, $x \neq y$, there exists a sequence $x = x_1, \ldots, x_n = y$ such that $x_i \in \mathrm{supp}(\xi)$, $1 \leq i \leq n$, and x_i is a neighbour of x_{i+1}, $1 \leq i \leq n-1$. If $\xi \in \Xi_0$ is not connected, clearly there exists a connected configuration $\xi_1 \in \Xi_0$ such that $C(\xi) >$

$C(\xi_1)$. Now let $\xi \in \Xi_0$ be a connected configuration being not a rectangular array. Let η_o be the smallest rectangular array in Z^d which contains supp (ξ). Define $\eta \in \Xi_0$ by supp $(\eta) = \eta_o$. It is easily seen that there exists an $x \in \eta_o$ - supp (ξ) such that x has least two neighbours $y, z \in$ supp (ξ). In this case, occupy x in some way and call the resulting configuration ξ . Clearly $C(\xi) > C(\xi')$. If $\xi' \neq \eta$, continue as before. This shows $C(\xi) > C(\eta) \geq 1$.

Concerning the first part of the second question, we refer to Doucet's question. Concerning the cell flow into the lattice, we refer to the skin model where exists a cell flow from the basal layer to the bottom, but this we did not examine. We have no information about other types of cell flows.

Dr. Steinijans:
Your simulation results are highly dependent on the transition probabilities. How did you obtain the transition probabilities, from the literature or by estimation from in vitro or in vivo data?

Tautu:
Our transition probabilities are chosen in agreement with some in vitro and in vivo data from the literature.

Concerning the paper of Rittgen and Tautu

O. Richter:
In general, the cell growth of a population of cells is a regulated process which is controlled by higher informational units. The model presented here is based on stochastic processes. Is it possible to extend the model to regulated processes ?

Rittgen:
The cell cycle model we presented here is in fact a stochastic controlled branching process.

Müller-Schauenburg:
Are all the parameters you are inserting into your model in vitro parameters, or do you have in vivo parameters as well ?

Rittgen:
Our parameters are chosen in agreement with some in vitro and in vivo data from the literature.

A Mathematical Model of Erythropoiesis in Man

H.E. Wichmann, H. Spechtmeyer, D. Gerecke and R. Gross

Medizinische Universitätsklinik, 5000 Köln, Germany
Supported by Grant DVM 301 from Bundesminister für Wissenschaft und Technologie

Introduction

The fundamental regulatory mechanisms of red blood cell formation which have been subject to extensive experimental research are nowadays well understood [1,2] . Thus, numerous attempts have been made to quantitative investigation of erythropoiesis by means of mathematical models. A number of models have been developed for single components of the erythropoietic system like the stem cell pool [3,4,5] , the erythropoietic cells in bone marrow [6,7,8] or the erythrocytes in blood [9,10] . Besides, several comprehensive models of erythropoiesis exist [11,12,13] . They have been verified by data from mammalians subjected to hypoxia [12] or treated with irradiation or iso-antibodies [11] .

The aim of this study is to describe the human erythropoiesis in a closed mathematical form and to compare the model results with data from healthy persons and from patients with anemic disorders.

The first chapter deals with the mathematical model: After a short medical survey (I.1) and a more detailed background on the biological facts and their mathematical formulation (I.2) the mathematical properties of the model (I.3) will be discussed. Definitions of diseases and courses of disease (I.4) follow and the chapter closes with numerical considerations (I.5). Chapter II compares model results and data measured in anemic patients (II.1) as well as in individuals at different altitudes (II.2). In the final discussion in chapter III the results will be interpreted and an outlook on further applications of the model will be given.

I The model

I.1. Medical survey [1]

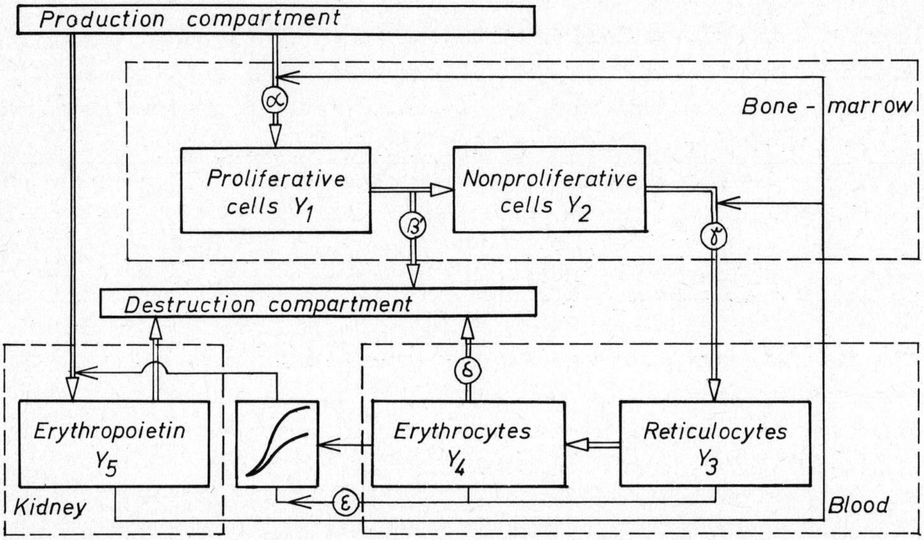

Figure 1 Model of the human erythropoiesis (⟹ Transitions of cells and hormones, → regulatory mechanisms, $\alpha, \ldots, \varepsilon$ disease parameters)

The block diagram of the model is shown in figure 1. The origin of the erythropoietic progeny is the stem cell pool, being situated in the production compartment. The stem cells develop to pronormoblasts, which represent the first stage of erythroid cells in the proliferation compartment Y_1. After some mitoses they switch to nonproliferative cells in compartment Y_2. Here they mature and extrude their nucleus. Afterwards, the cells emerge from the bone marrow and appear in the reticulocyte compartment Y_3 in the blood. They complete their maturation and finally transform to erythrocytes. These adult erythropoietic cells have a mean life-span of 120 days. Erythrocytes plus reticulocytes, in the following called red cells, have the function of oxygen transport.

The production of erythropoietic cells in the bone marrow is controlled by a hormone called erythropoietin, which influences the initial production rate and the maturation time. Erythropoietin production is regulated by the venous partial oxygen pressure of the kidney and thus indirectly by the number of red cells in blood. As symbolized in figure 1 on the left side of the erythrocyte compartment the position of the hemoglobin dissociation curve must be considered.

I.2 Biological facts and their mathematical formulation

On the basis of the compartment structure in figure 1 a more detailed discussion of the biological phenomena under consideration and their translation into mathematical expressions follows. Hence, we will restrict our attention to the 5 compartments $Y_1,\ldots,Y_5$ and regard the production and destruction compartments as the environment of the model, whose internal structure can be neglected. The erythropoietic disorders will be described by 'disease parameters' $\alpha,\ldots,\varepsilon$ with normal values $\alpha_N,\ldots,\varepsilon_N$ for healthy individuals.

Proliferative compartment Y_1

The initial rate of erythropoetic cells coming into the proliferative compartment consists of two parts [1] : A constant basic rate A' and an erythropoietin-dependent rate, for which the mathematical form $A''(1-e^{-BY_5})$ is assumed. Obviously,

$$A''(1-e^{-BY_5}) \rightarrow \begin{cases} 0, & \text{if } Y_5 \rightarrow 0 \\ A'', & \text{if } Y_5 \rightarrow \infty \end{cases}$$

The number of cells produced by mitosis is proportional to the number Y_1 of cells in the compartment: $C'\,Y_1$. The same holds for the cells leaving the compartment: $C''\,Y_1$. Disorders influencing this compartment are called aplastic anemias and defined by $\alpha < \alpha_N$. They are recognized by a diminished initial production rate of cells and caused by defects in the stem cell pool.

$$\dot{Y}_1 = \alpha(A-e^{-BY_5}) - CY_1 \qquad (1)$$

$$\alpha(A-e^{-BY_5}) = A'+A''(1-e^{-BY_5}),\ C = C'' - C'$$

A'	erythropoietin-independent initial production rate
$A''(1-e^{-BY_5})$	erythropoietin-dependent initial production rate
$C'Y_1$	proliferation rate
$C''Y_1$	flux into the nonproliferative compartment Y_2
$\alpha < \alpha_N$	disease parameter (aplastic anemias)

Nonproliferative compartment Y_2

Under physiological conditions all proliferative cells pass into the nonproliferative pool [1] : $C''Y_1$. The output of this compartment is equal to $G - e^{-HY_5}$. This term includes the influences of erythropoietin on the maturation time in the bone marrow [1] and has the property

$$G - e^{-HY_5} \rightarrow \begin{cases} G-1, \text{ if } Y_5 \rightarrow 0 & \text{(maximum maturation time)} \\ G, \text{ if } Y_5 \rightarrow \infty & \text{(minimum maturation time).} \end{cases}$$

Erythropoietic disorders can affect the nonproliferative compartment in two manners: First a destruction of marrow cells may occur (e.g. in pernicious anemia), called ineffective erythropoiesis and being regulated by the disease parameter $\beta' > \beta'_N$. Second a prolonged maturation time may appear, as observed in deficiency anemias and defined by $\gamma < \gamma_N$.

$$\dot{Y}_2 = \beta Y_1 - Z_6 Y_2, \quad \beta = C'' - \beta', \quad Z_6 = \gamma(G - e^{-HY_5}) \qquad (2)$$

$C''Y_1$	flux from the proliferative compartment Y_1
Z_6Y_2	flux into the reticulocyte compartment Y_3, including the erythropoietin dependence of the maturation time in the bone marrow
$\beta' > \beta'_N = 0$	ineffective erythropoiesis (e.g. pernicious anemia)
$\gamma < \gamma_N$	prolonged maturation time (deficiency anemias)

Reticulocyte compartment Y_3

All cells leaving the nonproliferative pool Y_2 enter the circulation as reticulocytes: Z_6Y_2. They leave this compartment as erythrocytes: Z_7Y_3. The sum of maturation times in Y_2 and Y_3 is constant [15] : $\frac{1}{Z_6} + \frac{1}{Z_7} = I$. Therefore, marrow cells which emerged early have to mature longer in blood, correspondingly.

$$\dot{Y}_3 = Z_6 Y_2 - Z_7 Y_3, \quad Z_7 = \frac{Z_6}{IZ_6 - 1} \qquad (3)$$

Z_6Y_2	flux from the nonproliferative compartment Y_2
Z_7Y_3	flux into the erythrocyte compartment Y_4
$\frac{1}{Z_6} + \frac{1}{Z_7} = I$	The sum of the maturation times in Y_2 and Y_3 is constant.

Erythrocyte compartment Y_4

All reticulocytes transform to erythrocytes: Z_7Y_3. The number of dying erythrocytes is assumed to be proportional to their total number: δY_4. The disease parameter $\delta > \delta_N$ regulates the increased destruction in hemolytic anemias.

$$\dot{Y}_4 = Z_7Y_3 - \delta Y_4 \tag{4}$$

Z_7Y_3	flux from the reticulocyte compartment Y_3
δY_4	destruction rate of erythrocytes
$\delta > \delta_N$	disease parameter (hemolytic anemias)

Erythropoietin compartment Y_5

The hormone production is regulated by the venous partial oxygen pressure Z_8 of the kidney [16,18] . Z_8 depends on the venous oxygen saturation according to the hemoglobin dissociation curve [2,14] and thus is related to the number of red cells. The relation between the production rate of erythropoietin and the partial pressure Z_8 is assumed to be exponential, as expressed by $D\,e^{-EZ_8}$. There is clinical evidence from observations in chronic anemias [1,17,28] , that this assumption is justified. The disappearance rate of the hormone corresponds to its concentration: FY_5. A disease parameter ε (usually called P_{50}) indicates the position of the hemoglobin dissociation curve [18] .In sickle cell anemia this curve is displaced to the right [24] : $\varepsilon > \varepsilon_N$.

$$\dot{Y}_5 = De^{-EZ_8} - FY_5, \quad Z_8 = \varepsilon\left[\frac{(Y_3+Y_4)^3 + J}{K(Y_3+Y_4)^3 - L}\right]^M \tag{5}$$

Z_8	venous partial oxygen pressure of the kidney (The formula is derived in detail in [48])
FY_5	destruction rate of erythropoietin
$\varepsilon > \varepsilon_N$	disease parameter (P_{50} in sickle cell anemia)

The model needs 13 constants A,...,M and 5 disease parameters $\alpha,\ldots,\varepsilon$. The values of A,...,M and the normal values $\alpha_N,\ldots,\varepsilon_N$ are taken from the literature. They are derived from cell numbers, maturation times, destruction rates and general dependences on the degree of anemia.

I.3 Mathematical properties

The 5 model equations (1) to (5) form a nonlinear system of differential equations:

$$\dot{Y} = f(Y), \ Y = (Y_1, \ldots, Y_5)' \tag{6}$$

It does not depend explicitly on the time and is called an autonomous system. With regard to its dependence on the disease parameters $\alpha, \ldots, \varepsilon$ it will be written as

$$\dot{Y} = f_P(Y), \quad P = (\alpha, \ldots, \varepsilon) \tag{7}$$

For every initial value

$$Y(t_0) = Y_0 \tag{8}$$

the system has a unique solution in the parameter region of interest. One can show that all partial derivates of f_P exist and are continuous and bounded. Thus the Lipschitz condition is fulfilled, which guarantees the existence of a unique solution $Y_p(t)$ of the initial value problem (7),(8) [19] . Moreover, the system has the property of asymptotic stability [20] : Every solution $Y_P(t)$ converges independent of the initial value Y_0 to an asymptotically stable value $\bar{Y}_P$, which depends on the parameter set P only:

$$\lim_{t \to \infty} Y_P(t) = \bar{Y}_P = \text{const.} \tag{9}$$

For a detailed discussion of these theoretical aspects see [48] .

I.4 Diseases, courses of disease

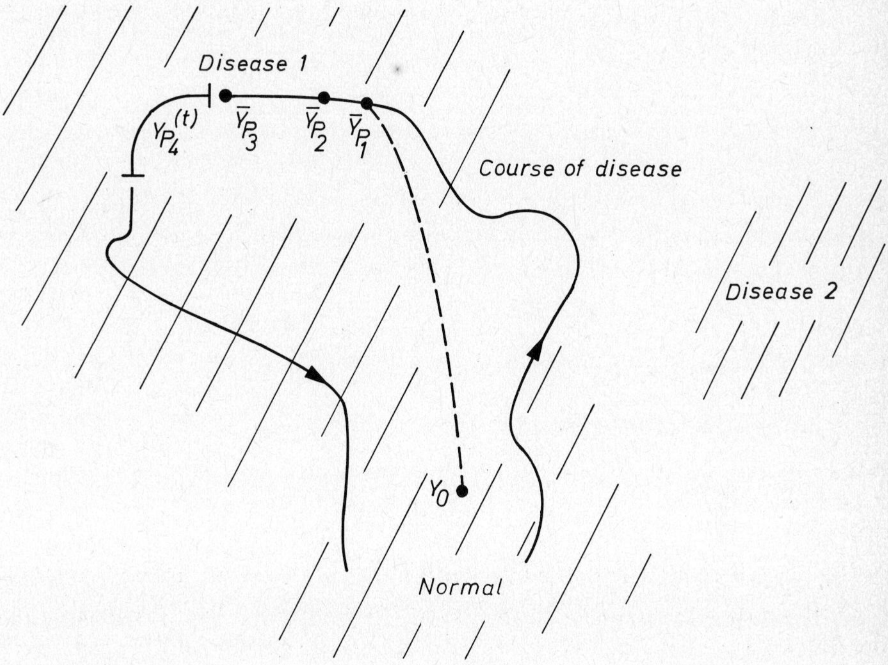

Figure 2 Diseases in the model. (—— course of disease, $\bar{Y}_{P_1}$, $\bar{Y}_{P_2}$, $\bar{Y}_{P_3}$ states of disease, $Y_{P_4}(t)$ course of therapy, – – – solution of the initial value problem $\dot{Y}=f_{P_1}(Y), Y(t_0)=Y_0$ with an arbitrarily chosen initial value Y_0 and the asymptotic stable state $\bar{Y}_{P_1}$)

The definitions of 'disease' and 'course of disease' are shown in figure 2: Diseases are regions in the space of the 5 variables $Y_1,\ldots,Y_5$, and a course of disease is representable by a path in this space. The development of an erythropoietic disorder can be interpreted in terms of the model as a gradual change of the disease parameters. The deviations of these parameters from the normal values indicate the degree of the disorder. Since the parameters change slowly compared with observation time, every state of disease can be described by a constant parameter set. Thus, the course of disease can be considered as a series of stable states of disease.

The parameter set P_1 in figure 2 corresponds to the state of disease $\bar{Y}_{P_1}$ which represents the asymptotic solution of the initial value problem $Y=f_{P_1}(Y)$, $Y(t_0)=Y_0$ with an arbitrarily chosen initial value Y_0.

The same holds true for $\bar{Y}_{P_2}$ and $\bar{Y}_{P_3}$. Likewise it is possible to simulate courses of therapy, if during the considered period the disease parameters can be regarded as constants. In figure 2 the state $\bar{Y}_{P_3}$ before therapy is taken as initial value, and the solution $Y_{P_4}(t)$ of $\dot{Y}=f_{P_4}(Y)$ shows the course of therapy.

In praxi the measured disease parameters $\hat{P}$ need not be optimal for the calculations and one has to find a set P of model parameters, which reproduces all experimental data satisfactorily. One possible way to find the optimal P is to minimize the quadratic form [46]

$$\Omega^s(P)= \sum_{i=1}^{n} a_i(\hat{P}_i - P_i)^2 + \sum_{i=1}^{m} b_i(\hat{\bar{Y}}_i - \bar{Y}_{P,i})^2 \tag{10}$$

for states of disease or

$$\Omega^c(P)= \sum_{i=1}^{n} a_i(\hat{P}_i - P_i)^2 + \sum_{j=1}^{k} c_j\Big(\sum_{i=1}^{m} b_i(\hat{Y}_i(t_j) - Y_{P,i}(t_j))^2\Big) \tag{11}$$

for courses of therapy with

the disease parameters $\hat{P}=(\hat{P}_1,\dots,\hat{P}_n)$ and $P=(P_1,\dots,P_n)=(\alpha,\dots,\varepsilon)$,
the states of disease $\hat{\bar{Y}}=(\hat{\bar{Y}}_1,\dots,\hat{\bar{Y}}_m)$ and $\bar{Y}_P=(\bar{Y}_{P,1},\dots,\bar{Y}_{P,m})$ and
the courses of therapy $\hat{Y}(t)=(\hat{Y}_1(t),\dots,\hat{Y}_m(t))$ and $Y_P(t)=(Y_{P,1}(t),\dots,Y_{P,m}(t))$

in measurement ('^') or model. The a_i, b_i, c_j are weighting factors, and the t_j are times of measurement. Hence, to express that not only the variable values $\hat{Y}$ and Y but also the disease parameters $\hat{P}$ and P will be compared, the pairs $(\hat{P},\hat{\bar{Y}})$ and $(P,\bar{Y}_P)$ will be called states of disease and $(\hat{P},\hat{Y}(t))$ and $(P,Y_P(t))$ will be called courses of therapy.

I.5 Numerical considerations

The numerical method used for the solution of the initial value problem (7), (8) is an extrapolation method [21,22]. The calculations were carried out on a CDC Cyber 72/76 and on a Siemens Unidata 7.730. The computer program needs 60 K Bytes, the calculation time per simulation takes only a few minutes. No serious numerical problems occured.

II Comparison of model results and data

II.1 Anemias

In the first section data from anemic patients will be presented, and states of disease as well as courses of therapy will be compared with results of model calculations. Each disease is described by at most 2 individually specified parameters. Both disease parameters and corresponding values of the model variables will be compared with the measured values. As already mentioned, these have been taken from medical literature, and they are therefore not always complete. Often only a part of the variables is known, and one or even both disease parameters may be missing. Nevertheless, these data can be used to verify the model. Several examples show that generally the replacement of free disease parameters by measured parameters of similar cases from other authors leads to a good reproduction of the rest of the data. Thus, even these unknown parameters cannot be chosen arbitrarily but are largely determined.

The limitation of available data also explains the choice of fixed model constants, neglecting the individual biological variability. As can be seen from the production capacity of the bone marrow, this may introduce considerable deviations: The maximal bone marrow production ranges individually from 6 to 10 times the normal production, but in the model it is taken rigidly as 7.5 times the normal value. Thus , supplementary measurements probably will lead to an additional improvement of the results.

In figure 3 states of disease from 8 patients with sickle cell anemia (SCA) or different hemolytic anemias (HA) are shown [23] . At first the way of presentation should be explained. The form of a wind rose has been chosen to compare the model results with the measured data. The patient's values are localized in the centre of the wind rose, and the model values are drawn in the different directions. The length of the arm is normalized to the accuracy of the experimental method of investigation. The large circle represents the 95% confidence area.

As can be read off the wind rose in figure 3, the variables Y_1+Y_2, Y_3 and Y_3+Y_4 have been measured by this author. They are increase of production in the bone marrow, reticulocytes and number of red cells. Furthermore, the erythrocyte life-span δ has been investigated. However, for sickle cell anemia (SCA) a second disease parameter ε is

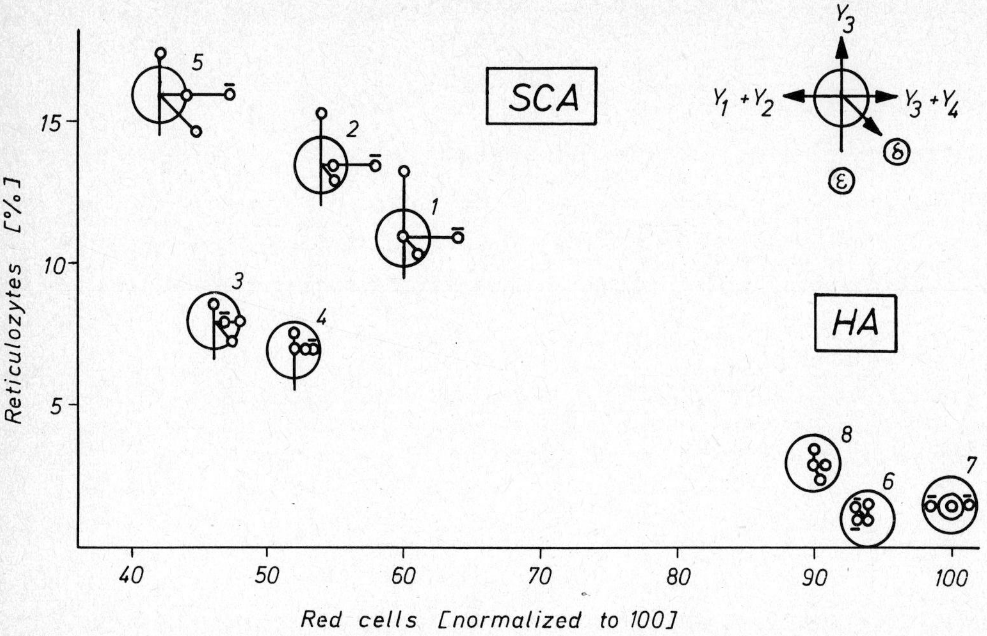

Figure 3 States of disease in sickle cell anemia (SCA) and other hemolytic anemias (HA). Comparison of model results and data from [23]. Large circle: 95% confidence area around the measured values localized in the centre. Small circles:Positions of the calculated values. The arms are normalized to the experimental accuracy.

necessary [24] . Its value is not available, and this is symbolized by deleting the arrow-head and the small circle. Finally, every case is localized in the reticulocyte-red cell-plane.

As figure 3 shows the model reproduces the data from the SCA-patients 3 and 4 within the accuracy of measurement: All small circles lie within the large circle. For the patients 1, 2 and 5 the calculated number of reticulocytes Y_3 is a little too large, and the increase of production in the bone marrow Y_1+Y_2 is somewhat too small, shown in the minus sign above the corresponding small circle.

In the cases of different hemolytic anemias (HA) measurements and calculations agree within the experimental accuracy. Here we have no free parameter, because the erythrocyte life-span δ is the only disease parameter for these disorders.

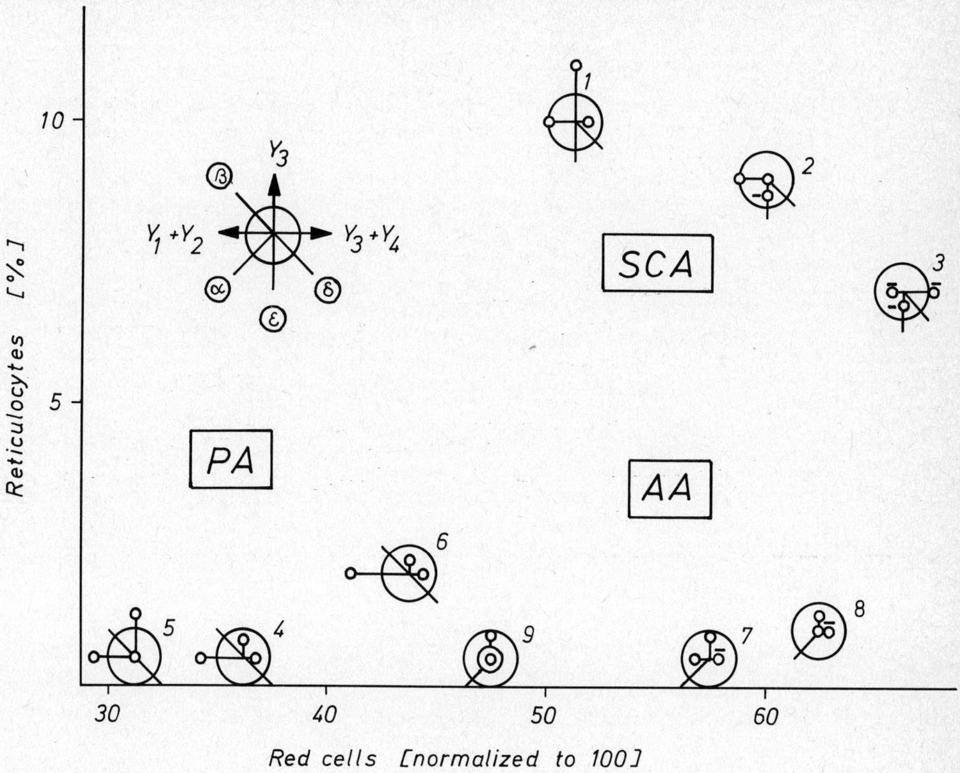

Figure 4 States of disease in sickle cell anemia (SCA), pernicious anemia (PA) and aplastic anemias (AA). Comparison of model results and data from [25] .

Figure 4 shows data from 3 different anemias [25] . The sickle cell anemia (SCA) with the disease parameters δ and ε is already known. For the aplastic anemias (AA), only the disease parameter α is changed, which regulates the initial production rate in the bone marrow. In pernicious anemia (PA) β and δ are altered, which means that bone marrow cells die and simultaneously the destruction rate of erythrocytes is enlarged. Additionally, pernicious anemia is a deficiency disease. This fact is taken into account by a global prolongation factor γ for the marrow maturation time.

The increase of production in the bone marrow Y_1+Y_2, reticulocytes Y_3 and red cells Y_3+Y_4 have been measured, informations about the disease parameters are not available. The results show that the model describes the 3 different diseases quite well.

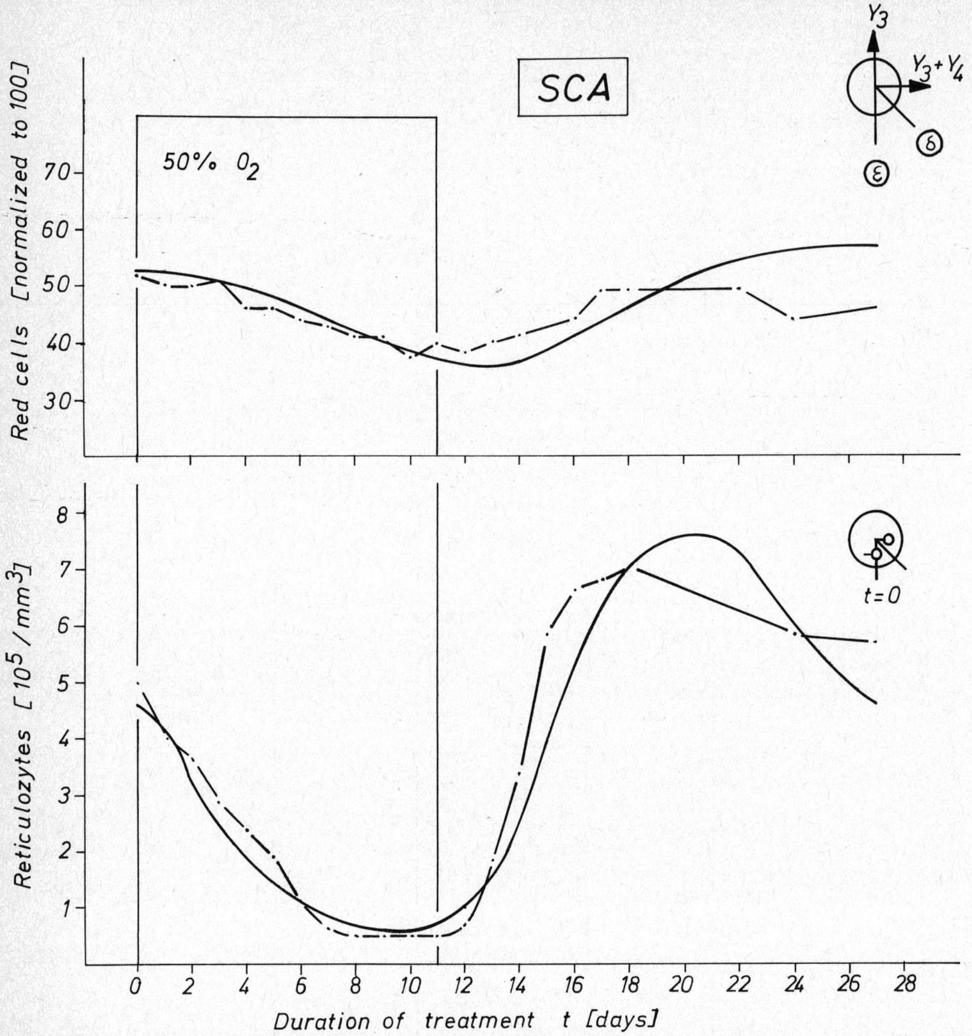

Figure 5 Sickle cell anemia (SCA) with oxygen therapy. Comparison of model results (———) and data from [26] (—·—).

The first course of therapy is presented in figure 5 [26] .A patient with sickle cell anemia (SCA) is treated with oxygen. According to the good oxygen supply of the tissues the erythropoietin production is suppressed, the bone marrow cell production nearly stops and the maturation time is prolonged. The observable effect in blood is a sharp reduction of the reticulocytes within a few days and a slower decrease of the red cell numbers. The decrease at the beginning as well as the increase after cessation of therapy are satisfactorily reproduced in both curves. The state of disease before therapy is shown on the right.

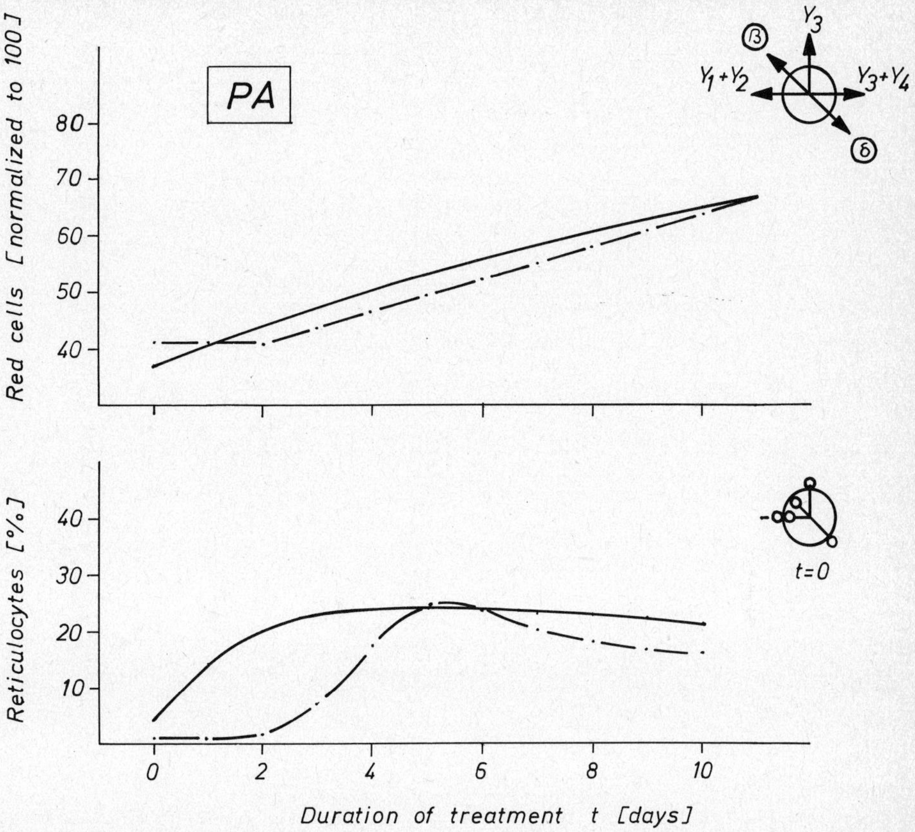

Figure 6 Pernicious anemia (PA) with vitamin B_{12}-therapy. Comparison of model results (———) and data from [27] (—·—).

Figure 6 presents data from a patient with pernicious anemia (PA) treated with vitamin B_{12} [27] . This therapy stops the destruction of bone marrow cells. It shows an effect within a few hours, but it influences only the formation of new cells. This causes a time delay for the increase of the measured reticulocyte curve. In the model this special property of the vitamin B_{12} treatment is not considered, and therefore the calculated reticulocyte curve rises at once. Apart from this timelag the curves agree as well as the disease parameters and cell numbers before therapy.

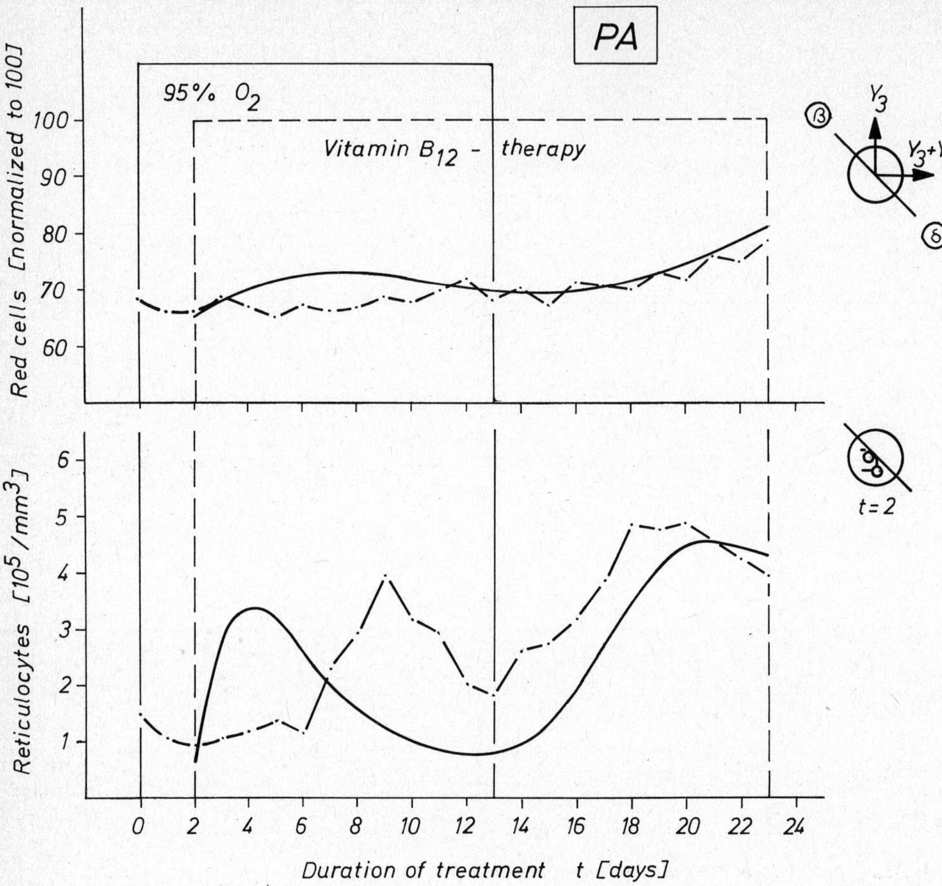

Figure 7 Pernicious anemia (PA) with combined oxygen and vitamin B_{12} therapy. Comparison of model results (——) and data from [26] (—·—).

The data of the last patient are shown in figure 7 [26]. He also suffers from pernicious anemia (PA) and is treated with oxygen and vitamin B_{12}. The vitamin therapy causes an increase of reticulocytes, but the simultaneous oxygen therapy makes them decrease soon. After termination of oxygen treatment reticulocytes rise again. Once more the time delay between both curves at the beginning of vitamin B_{12}-therapy is to be seen. Curves and states of disease in model and measurement agree satisfactorily.

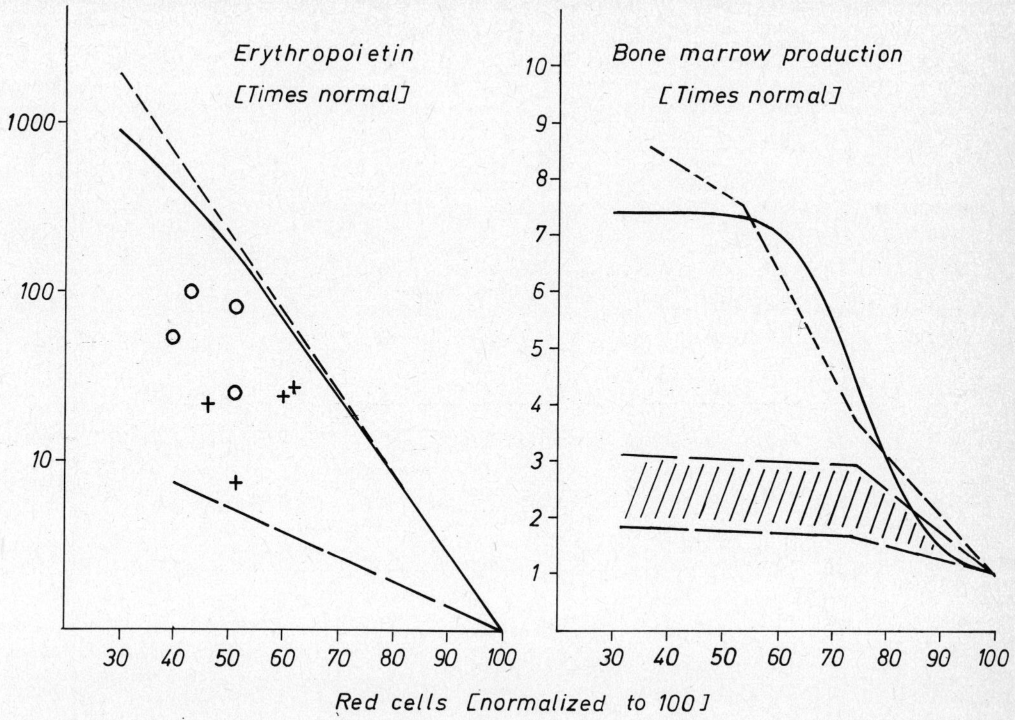

Figure 8 Erythropoietin and bone marrow production depending on the degree of anemia. Left: Comparison of model results for anemias without P_{50}-displacement (———) and some arbitrary cases of sickle cell anemia (+) with data from [17] (– – –, o). Right: Comparison of model results (———) and data from [1] (– – –) for anemias without proliferation disorders. Different curves for acute anemias are given by [28] (—— —— left) and [1] (—— —— right).

Two final diagrams show general effects of anemias. In figure 8 the relation between the degree of anemia and erythropoietin [17] or marrow production [1] is demonstrated. The presented curves agree well. However, the curves should be interpreted qualitatively only. The experimental erythropoietin values have a large variability - the drawn curve is a regression line - , and the dependence of the marrow production on the number of red cells shows remarkable individual differences, as discussed in section II.1.

The curves for acute anemias [28,1] clearly differ from those of the chronic forms and show the limits of the model: Until now the complicated effects of iron deficiency in these disorders cannot be reproduced.

II.2 Effects of altitude

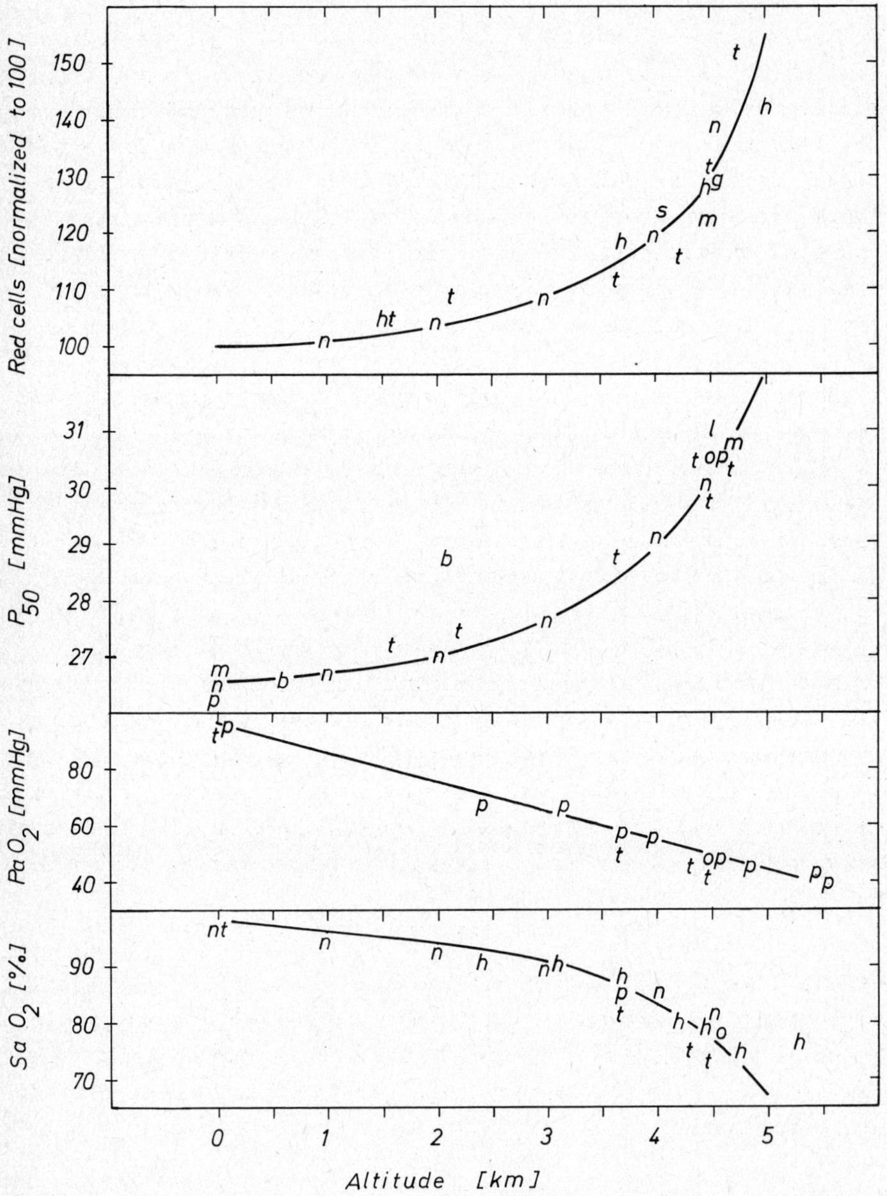

Figure 9 Red cells, P_{50} and arterial PO_2 and SO_2 in residents at different altitudes. Comparison of model results (2 ——) and data from several authors [29-44] .

The second part of the data concerns the regulatory system of healthy individuals at moderate and high altitudes. In figure 9 the dependence of red cells and some oxygen variables on altitude is shown. The letters represent the results of different publications [29 - 44] ,every point is an average value of several persons.

As can be seen from the uppermost curve in figure 9, the red cell numbers in residents rise slowly from 100 at sea level to 110 at 3 km altitude. Above 4 km the increase is steep and reaches the value of 150 at 5 km altitude. This region seems to be the upper limit for human settlements [1] . Similar to the red cells the P_{50} of the hemoglobin dissociation curve increases nearly exponentially. The decrease of the arterial partial oxygen pressure PaO_2 can be represented by a line, and the arterial oxygen saturation SaO_2 falls in a negative exponential way. The model curves reproduce the data well.

After these steady state curves in figure 9, adaptation phenomena in subjects moving from sea level to high altitudes are shown in figure 10 [29 - 44] . The erythropoietin curve of the model reaches its peak within 3 days, then it decreases slowly to twice the normal value after 100 days. The measured erythropoietin results increase similar to the model curve, but they clearly decrease earlier. Better agreement is found for the cell numbers. Both curves and points of measurement show a maximum value within 3 to 5 days for the bone marrow cells and within 7 to 10 days for the reticulocytes. The curves return to nearly normal value after 100 days, when the steady state is reached. Red cells behave differently: They increase monotoneously from 100 to 130, which is the value for residents living at the considered altitude of 4.5 km. Curves and data agree well, especially when the confidence areas for erythropoietin and reticulocyte measurements are taken into account.

The steady states from residents, shown in figure 9, have to be interpreted in terms of the model as states of disease, the adaptation processes in figure 10 correspond to courses of therapy. In both cases arterial PO_2 and P_{50} of the considered altitude have been used as disease parameters.

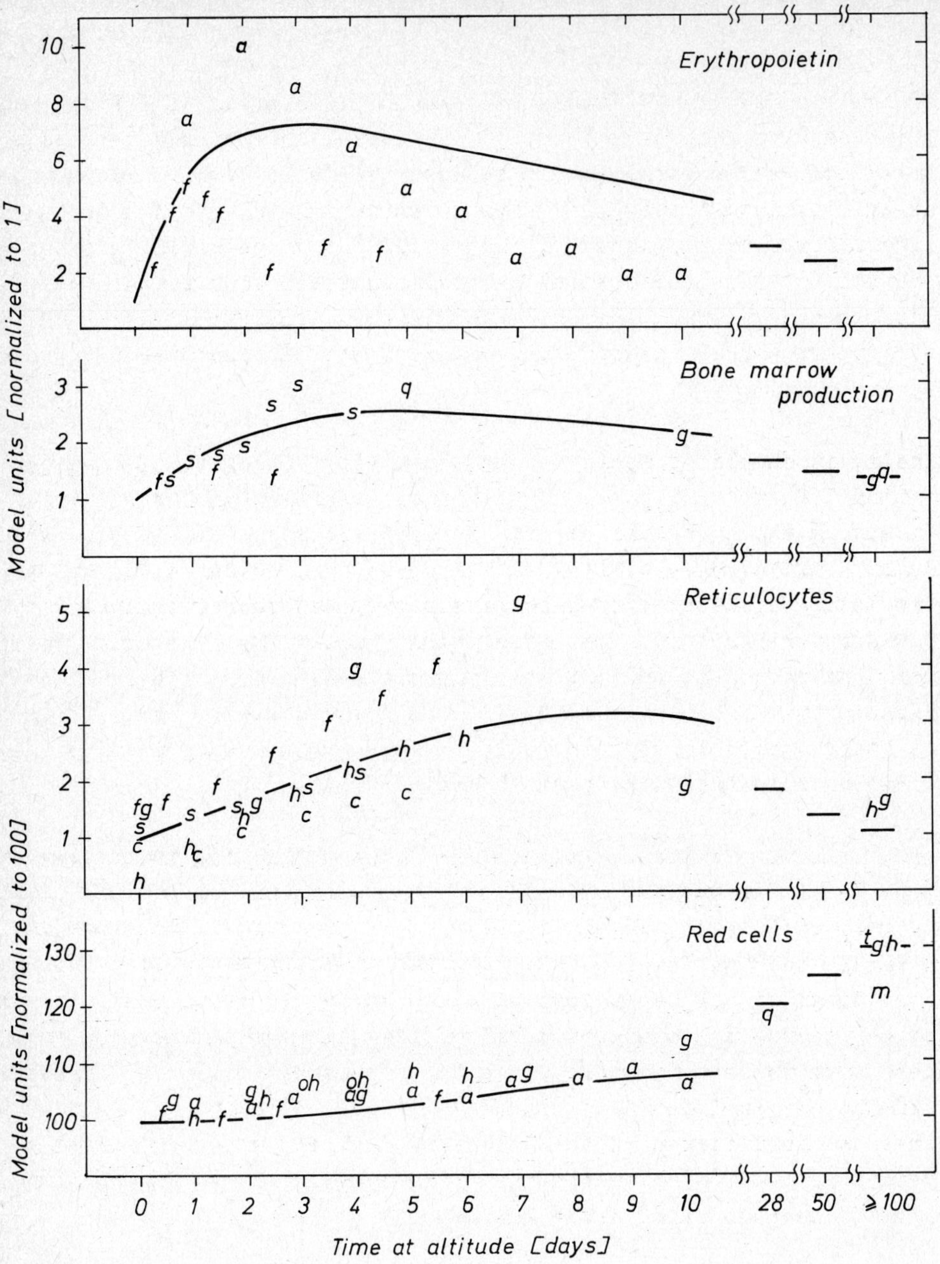

Figure 10 Changes in erythropoiesis after ascent from sea level to 4.5 km altitude. Comparison of model results (———) and data from several authors [29-44] .

III Discussion

In total, more than 60 states of disease, 10 courses of therapy and 11 curves with general dependences on anemia or altitude have been simulated. Such a large amount of data could be obtained from the medical literature only, and thus incomplete data sets had to be accepted. However, the results, only a part of which could be presented here [48] , allow one assertion: It is possible to describe the complex human erythropoiesis by a simplified mathematical model and to reproduce measured values from very different influences on the regulatory system.

Finally it should be mentioned that not all 'biological facts' in section I.2 are undisputed. There exist contrary hypotheses, e.g. concerning the questions of erythropoietin consumption in the bone marrow [45,1] , of extrarenal erythropoietin production [1] or of stimulating effects on erythropoiesis by the destruction products of the red cells [1] . This study merely shows that the model assumptions do not contradict the avaiable measurements, but one cannot conclude that the corresponding medical hypotheses are verified. On the other hand, it is possible to exclude wrong assumptions, if contradictions to the experimental data occur [46,3,8] .

Thus, a first practical application of the model lies in the test of hypotheses. Furthermore it can be used to simulate therapies or other external influences on erythropoiesis, if the effects of these influences can be quantified. A third possible application is the computer diagnosis of hematological diseases. With the help of the model one can reduce the measured data to a few disease parameters , and these seem to be more appropriate to the classification of diseases than the original data. Additionally, model calculations could be a first toe hold for an adequate mathematical analysis of courses of therapy or even the development of diseases, problems for which no suitable methods exist until now.

Summary

In this paper a model of red cell formation in man is presented. Mathematically it is formulated as a compartment model and consists of five coupled nonlinear differential equations. It describes the normal as well as the pathological erythropoietic regulatory system and reproduces states of disease as well as courses of measurements under therapy. Besides data from patients with different anemias (aplastic anemias, pernicious anemia or sickle cell anemia) data from individuals at moderate and high altitudes are compared with the results of model calculations. The considered data which have been taken from the medical literature include investigations of more than 60 anemic patients and some hundred healthy persons. In nearly all cases a good quantitative agreement between calculation and measurement has been found.

References

1 Williams,W.J.,Beutler,E.,Erslev,A.J.,Rundles,R.W.: Hematology, McGraw-Hill New York (1972)

2 Ganong,W.F.: Lehrbuch der med. Physiologie. Springer Berlin (1974)

3 Lajtha,L.G.,Oliver,R.,Gurney,C.W. Brit.J.Haemat. 8 (1962) 442-460

4 Newton,C.M. Ann.N.Y.Acad.Sci. 128 (1966) 781-789

5 Kretchmar,A.L. Science N.Y. 152 (1966) 367-370

6 Tarbutt,R.G.,Blackett,N.M. Cell Tissue Kinet. 1 (1968) 65-80

7 Lala,P.K.,Patt,H.M.,Maloney,M.A. Acta Heamt. 35 (1966) 311-318

8 Hanna,I.R.A.,Tarbutt,R.G. Cell Tissue Kinet. 4 (1971) 47-59

9 Zajicek,G. J.Theor.Biol. 19 (1968) 51-66

10 Garby,L.,Groth,T.,Schneider,W. Comp.and Biochem.Res.2 (1969) 229

11 Kirk,J.,Orr,J.S.,Hope,C.S. Brit.J.Haemat.15 (1968) 35-47

12 Mylrea,K.C.,Abbrecht,P.H. J.Theor.Biol. 33 (1971) 279-297

13 Duechting,W. Blut 27 (1973) 342-350

14 Hill,A.V. J.Physiol. 40 (1910) IV-V

15 Hillman,R.S. J.Clin.Invest. 48 (1969) 443-453

16 Schulz,E. Internist (Berlin) 12 (1971) 210-215

17 Alexanian,R. J.Lab.Clin.Med. 82 (1973) 438-445

18 Metcalfe,J.,Dhindsa,D.S. from Rorth,M.,Astrup,P.:Oxygen Affinity of Hemoglobin and Red Cell Acid Base Status. Munksg.Copenh.(1972)

19 Walter,W.:Gewöhnliche Differentialgl.HT 110.Springer Heidelb.(1972)

20 Cesari,L.:Asymptotic Behavior and Stability Problems in Ordinary Differential Equations. Springer Berlin (1959)

21 Stör,J.,Bulirsch,R.: Einf.i.d.num.Math.II.HT 114.Springer (1973)

22 Bulirsch,R.,Stör,J. Num.Math. 8 (1966) 1-13

23 McCurdy,P.R. Blood 33 (1969) 214-224

24 Milner,F.P. Arch.Intern.Med. 133 (1974) 565-572

25 Bothwell,T.H.,Hurtado,A.V.,Donohue,D.M.,Finch,C.A. Blood 12 (1957) 409-427

26 Tinsley,J.C.,Moore,C.V.,Dubach,R.,Minnich,V.,Grinstein,M. J. Clin. Invest. 28 (1949) 1544-1564

27 Hillman, R.S.,Adamson,J.,Burka,E. Blood 31 (1968) 419-433

28 Adamson,J.W. Blood 32 (1968) 597-609

29 a Abbrecht,P.H.,Littell,J.K. J.Appl.Physiol. 32 (1972) 54-58

30 b Amor,H.,Humpeler ,E.,Deetjen,P. Wien.Klin.Wochenschr.85(1973)700

31 c Carmena,A.O.,Testa,N.G.,Frias,L.F. Proc.Soc.Exp.Biol.Med. 125 (1967) 441-443

32 f Faura,J.,Ramos,J.,Reynafarje,C.,English,E.,Finne,P.,Finch,C.A. Blood 33 (1969) 668-676

33 g Huff,R.L.,Lawrence,J.H.,Siri,W.E.,Wasserman,L.R.,Hennessy,T.G. Medicine 30 (1951) 197-217

34 h Hurtado,A.,Merino,C.,Delgado,E. Arch.Intern.Med. 75 (1945) 284-323

35 l Lenfant,C.,Torrance,J.,English,E.,Finch,C.A.,Reynafarje,C.,Ramos, J.,Faura,J. J.Clin.Invest. 47 (1968) 2652-2656

36 m Lenfant,C.,Ways,P.,Aucutt,C.,Cruz,J. Resp.Physiol. 7 (1969) 7-29

37 n Lenfant,C.,Torrance,J.D.,Woodson,R.,Finch,C.A. from Brewer,G.J.: Advances in Experimental Medicine and Biology,Vol. 6. Plenum Press. New York (1970)

38 o Lenfant,C.,Torrance,J.D.,Reynafarje,C. J.Appl.Physiol.30(1971)625

39 p Pace,N. Fed.Proc.33 (1974) 2126-2132

40 q Reynafarje,C.,Villavicencio,D.,Faura,J. Acta Physiol.Lat.Am. 23 (1973) 134-135

41 r Roerth,M.,Nygaard,S.F.,Parving,H.H.,Hansen,V.,Kalsig,T. Scan.J. Clin.Lab.Invest. 31 (1973) 447-452

42 s Siri,W.E.,Van Dyke,D.S.,Winchell,H.S.,Pollycove,M.,Parker,H.G., Cleveland,A.S. J.Appl.Physiol. 21 (1966) 73-80

43 t Torrance,J.D.,Lenfant,C.,Cruz,J.,Marticorena,E. Resp.Physiol. 11 (1970-71) 1-15

44 z Scaro.,J.L.,Guidi,E.E. Acta Physiol.Lat.Am. 20 (1970) 281-283

45 Essers,U.,Mann,H. Blut 22 (1971) 297-304

46 Sendov,B.,Tsanev,R. from Bailey,N.T.J.,Sendov,B.,Tsanev,R.:Mathematical Models in Biology and Medicine. North Holland/American Elsevier . Amsterdam (1974)

47 Monot,C.,Martin,J. from Bailey,N.T.J.,Sendov,B.,Tsanev,R.:Mathematical Models in Biology and Medicine. North Holland/American Elsevier . Amsterdam (1974)

48 Wichmann,H.E.: Dissertation (in preparation)

Simulation of Biochemical Pathways and its Application to Biology and Medicine

Otto Richter and Augustin Betz

Medizinische Einrichtungen der Universität Düsseldorf
Institut für Medizinische Statistik und Biomathematik
Moorenstraße 5, 4000 Düsseldorf

Botanisches Institut der Universität Bonn
Kirschallee 1, 5300 Bonn

Summary

In recent years quantitative data on biochemical pathways have accumulated up to a point where a detailed mathematical analysis of multienzymesystems concerning their dynamics and regulatory properties has become possible: the knowledge of the structure of a pathway, its metabolite scheme, the pool sizes involved, and the kinetic laws of its enzymes enable its mathematical description in terms of coupled non-linear differential equations.

In the following paper the general form of the basic kinetic equations containing all species of a multienzymesystem is represented. Assuming a stationary state hypothesis for the enzyme species these equations are reduced to a system of conservation equations for the fluxes of substrates and intermediates whose terms consist of the kinetic laws of the corresponding enzymes.

Two examples of application in biology and medicine are given: simulation of glycolytic oscillations and simulation of allopurinol-therapy in hyperuricemia, e. d., the action of allopurinol on the last two steps in purine degradation.

List of abbreviations

HK:	Hexokinase (E.C. 2.7.1.1)
PFK:	Phosphofructokinase (E.C. 2.7.1.11)
F6P:	Fructose-6-Phosphate
FDP:	Fructosediphosphate
ATP, ADP, AMP:	Adenosin (tri, di, mono) phosphate
XANT:	Xanthine
HYPO:	Hypoxanthine
v_x:	kinetic law of enzyme x
AllO:	Allopurinol

Contents

I Theoretical background

1. Biochemical reaction systems

Biochemical reaction systems are composed of metabolites $s_i(t)$ $(i = 1, ..., k)$ and enzymes e_j $(j = 1, ..., l)$ in various forms like free enzymes, enzyme-substrate complexes and enzyme-effector complexes. The kinetics of a multienzymesystem is described by a set of coupled non-linear differential equations of first order with respect to time [1] :

$$\frac{d}{dt} s_i(t) = \sum_{j=1}^{k} S_{ij}(s_1,...,s_k,k_1,...,k_r)\, e_j + c_i(s_1,...,s_k, k_1,...,k_r)$$

$$\frac{d}{dt} e_j(t) = \sum_{n=1}^{l} E_{jn}(s_1,...,s_k,k_1,...,k_r)\, e_n + d_j(s_1,...,s_k, k_1,...,k_r)$$

(1a,1b)

where $k_1,...,k_r$ denote the microscopic rate constants. S_{ij}, E_{jn}, c_i, d_j are linear functions of the microscopic constants and the concentrations of substrates and enzyme species. The conservation laws for the s_i and e_j are incorporated into the system, so the variables are linear independent.

In general, the system cannot be handled as it stands: (i) a tremendous number of microscopic rate constants has to be fed into the system. Only very few of these rate constants have been measured so far.

(ii) the large number of equations renders the system unintelligible. The interpretation of numerical results obtained by simulation may be more difficult as the interpretation of experimental results.

(iii) the system contains information on the time course of the enzyme species, which cannot be followed experimentally. Experimental data of multi-enzyme systems are given in terms of pool sizes, metabolite concentrations and macroscopic kinetic constants of the kinetic laws of the enzymes.

2. Steady state approximation

The system is heterogen with respect to concentration ranges. For many enzyme systems the following inequations hold

$$c_{enzyme} \ll c_{metabolite} < c_{pool} \qquad (2)$$

where $\ll$ denotes a factor of at least one order of magnitude in the ratio $c_{enzyme}/\,c_{metabolite}$. The heterogen concentration ranges are related to the time hierarchy of the system. With respect to the metabolic time τ_m [2] the motion of a subsystem is defined to be slow if

$$\frac{\tau_m}{\tau_s} \ll 1 \qquad (3)$$

and fast if

$$\frac{\tau_m}{\tau_f} \gg 1 \qquad (4)$$

If $c_{enzyme} \ll c_{metabolite}$ holds the motion of the enzyme subsystem is fast with respect to the metabolite subsystem, and the ratio of the characteristic times is approximately given by

$$\frac{\tau_f}{\tau_m} \approx \frac{c_{enzyme}}{c_{metabolite}} \qquad (5)$$

In this case the Tikhanov theorem [3] can be applied to the equation system (1a,1b): the time derivatives of the enzyme species are set equal to zero, and the linear equations for the enzyme subsystem can be solved. Equations (1a,1b) written in vector notation then become

$$\frac{d}{dt}\vec{s} = S\,\vec{s} + \vec{c} \qquad (6a)$$

$$\frac{d}{dt}\vec{e} = E\,\vec{e} + \vec{d} = 0 \qquad (6b)$$

From equation (6) $\vec{e}$ can be solved in terms of $\vec{s}$.

$$\vec{e} = -E^{-1}\vec{d} \qquad (7)$$

So equations (6) can be written in the following form:

$$\frac{d}{dt} s_i(t) = V_i - V_{-i} \tag{8}$$

Where V_i and V_{-i} are net rate laws governing the production and consumption of the metabolites or intermediates s_i.

For a linear enzymatic chain

$$\xrightarrow{v_o} s_o \xrightarrow{v_1} s_1 \longrightarrow \cdots\ s_n \xrightarrow{v_n}$$

the system becomes

$$\frac{d}{dt} s_i(t) = v_{i-1}(s_1,\ldots,s_n,\ K_1,\ldots,K_r) - v_i(s_1,\ldots,s_n,K_1,\ldots,K_r) \tag{9}$$

where v_i is the kinetic law of the i th enzyme with the macroscopic kinetic constants $K_1,\ldots,K_r$.

The kinetic laws and the corresponding constants are known for many enzymes by in vitro studies and can be inserted into the equation system (8). Equations (8) are thus a suitable basis for the analysis and simulation of biochemical reaction systems, since they depend on macroscopic constants, which are known or which are in principle measurable. Furthermore, the time course of the metabolites can be checked by experiments. For the analysis and simulation the following information is required:

(i) the structure of the pathway
(ii) the kinetic laws of its enzymes
(iii) the pool sizes of cofactors as the adenylates or the NADH/NAD system

3. Mathematical Analysis vs Simulation

The kinetic laws $v_i(s_1,\ldots,s_n,K_1,\ldots,K_r)$ are in general non linear functions of substrate, product and effector concentrations, hence the resulting differential equations are non linear and cannot be solved explicitly. Nonetheless the system can be attacked by analytical means for the investigation of stability, limit cycle behaviour and other dynamic properties, which provide criteria in terms of kinetic constants and pool sizes. Unfortunately, the equations describing real systems are so complex that the analytical approach is tremendously difficult. An insight into the dynamics of such a system can be gained only by simulation, that means solving the system numerically and performing a lot of computer runs under different conditions as regards pool sizes, infusion rates or application of chemical pulses. The analytical approach is most useful for small model systems containing only few of the basic elements, for instance a linear chain with only one regulatory enzyme. The analytical investigation of those systems reveals interesting properties : a linear chain with negative feedback for instance exhibits limit cycle behavior under certain conditions [4]. The purpose of these models is to serve as a guide for the interpretation of the properties of larger systems which can be studied only by simulation.

II Application to Biology: Simulation of Glycolytic Oscillations

1. Biochemical background

Glycolytic oscillations have been observed in a variety of organisms and preparations, in whole yeast cells [5] and in cell extracts from various organisms as rabbit muscle [6], beaf heart [7] or yeast [8]. When glucose is added to the glycolytic system under anaerobic conditions the system starts oscillating: all concentrations of intermediates and cofactors of glycolysis change periodically in time. Whereas the metabolite pattern and frequencies vary with temperature infusion rate, and material, the phase relationships between intermediates and cofactors are always the same: F6P and FDP, the substrate and product of the key enzyme phosphofructokinase (PFK) are oscillating with a phase difference of nearly 180°, the phases of ATP and ADP differ by 180°, whereas AMP oscillates in phase with ADP. Similar relations hold for other intermediates and cofactors, so the phase pattern of oscillating glycolysis can be regarded as an intrinsic property of the system reflecting its regulatory structure. Since the component structure of glycolysis is very well known [9], this system is best suited for mathematical analysis and simulation.

2. The essential subsystem

There is much evidence that PFK with its complex allosteric behavior and the adenylate system (ATP, ADP, AMP) interact by positive and negative feedback (AMP activates PFK, ATP inhibits PFK) and that this interaction causes the oscillatory response to the addition of substrate. This is confirmed by analytical investigations of simplified model systems containing only PFK and one of the adenylates [10], [11]. Experimental results - in cell extracts oscillations can be initiated only by infusing F6P or one of its precursors [8] - support this hypothesis, although other mechanisms were discussed. So the PFK adenylate system is a suitable subsystem for studying the dynamics of glycolysis.

3. The Model

The essential subsystem is described by the reaction scheme presented in Fig. 1: it contains all essential elements of the glycolytic pathway, consumption of sugar, production of the energy rich compound ATP, and a sink for ATP, which is either consumed by storage of carbohydrates or by other processes. Furthermore, it contains all three adenylates ATP, ADP and AMP, so the energy charge EC = (ATP + 0.5 ADP)/(ATP + ADP + AMP) can be computed. Goldbeter [11] also computes the energy charge out of his model, omitting AMP, however.

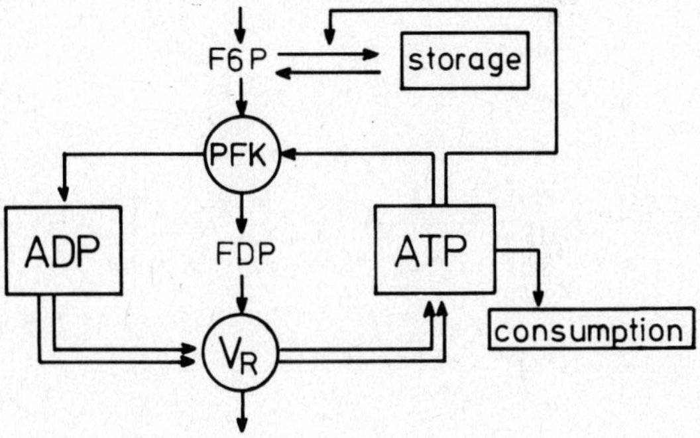

Fig. 1 Simplified reaction schema of glycolysis

This reaction scheme is put into equations according to the equation system (9):

$$\dot{[F6P]} = v_o - v_{PFK} - v_S \quad (10)$$

$$\dot{[FDP]} = v_{PFK} - v_R \quad (11)$$

$$\dot{[ATP]} = N\, v_R - v_{PFK} - v_S - v_{SINK} - k_1 [ATP]\,[AMP] + k_2 [ADP]^2 \quad (12)$$

$$\dot{[ADP]} = v_{PFK} + v_S + v_{SINK} + 2\, k_1 [ATP]\,[AMP] - 2\, k_2 [ADP]^2 - N\, v_R \quad (13)$$

$$[ADP] + [AMP] + [ATP] = const = N_o \quad (14)$$

$$v_{PFK} = \frac{V_{MAX}\, \overline{F6P}\, (\overline{F6P} + 1)^3}{\left(\frac{\overline{ATP} + 1}{\overline{AMP} + 1}\right)^4 + (\overline{F6P} + 1)^4} \; \frac{[ATP]}{K_m + [ATP]} \quad (15)$$

$$v_R = \frac{V_{MAX} [FDP]}{K_m + [FDP]} \quad (16)$$

$$v_{SINK} = k_{SINK} [ATP] \quad (17)$$

$$v_S = k_S [ATP]\,[F6P] \quad (18)$$

with $\overline{F6P} = [F6P] / k_{F6P}$, $\overline{ATP} = [ATP] / k_{ATP}$ and

$\overline{AMP} = [AMP] / k_{AMP}$

The interconversion of adenylates according to 2 ADP $\rightleftharpoons$ ATP + AMP is described by chemical rate equations. The term v_R, which stands for the lower part of glycolysis, is presented by a Michaelis -Menten law, so the model contains only one feedback.

The substrate (here F6P) is infused with constant velocity v_o into the system and is transformed to FDP in the PFK reaction. N units of ATP are produced in the lower part of glycolysis or additionally in the respiratory chain with the rate v_R. ATP is consumed with the rates v_{SINK} (consumption not coupled with glycolysis) and v_S (consumption by the storage of carbohydrates). The adenylates are interconverted in the adenylate kinase reaction: since two molecules of ADP form ohne molecule of ATP and one molecule of AMP, two equations are needed for the adenylate system, provided the sum of the adenylates remains constant in time. This assumption ist not always fulfilled in vivo [6].

For PFK a Monod model is taken (eq. 15) with homotropic effects by the substrate F6P, inhibition by ATP and activation by AMP. The role of ATP as cosubstrate is expressed by the factor ATP / (ATP + K_m). Since K_m is very low this factor is only important in situations where the ATP concentration is very low, too. The constants used in the model are given in table 1.

The system is solved numerically by the subroutine HPCG [12] which is based on Hammings predictor corrector method. The program is conceived to allow a dialog with the user, who can choose alternative kinetic laws and constants andwho is able to run the program without any knowledge in programming.

4. Results

The time evolution of the adenylates and the energy charge was tested in several computer runs under a variety of conditions. The computer results are compared with experimental curves obtained in yeast cells.

Fig. 2 shows the response of the adenylates in the model to the addition of glucose under anaerobic conditions, that means, only two units of ATP are produced for the breakdown of one unit glucose.

Because ATP is consumed first in the PFK reaction, its concentration decreases, whereas the concentration of AMP increases. ADP follows ATP. The relative long delay before the onset of ATP production is characteristic for starved yeast cells.

Under anaerobic conditions, where 38 units of ATP are produced the ATP drop is much shorter: in this case ADP follows AMP.

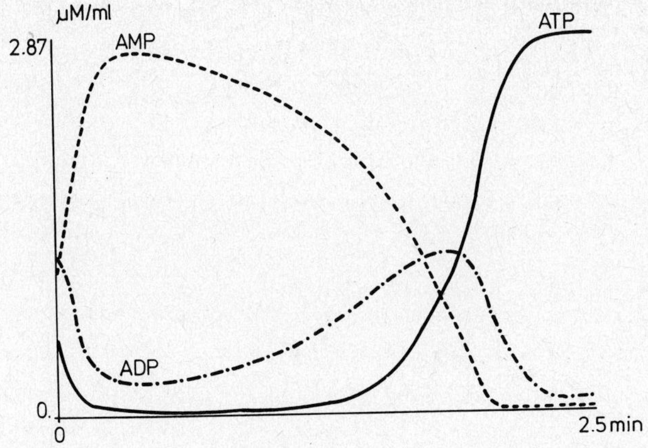

Fig. 2 Response of the adenylates in the model system to the addition of glucose (unaerobic conditions)

The experimental progress curves of the adenylates as published by Kopperschläger et. al [13] are in agreement with the model curves (Fig. 3).

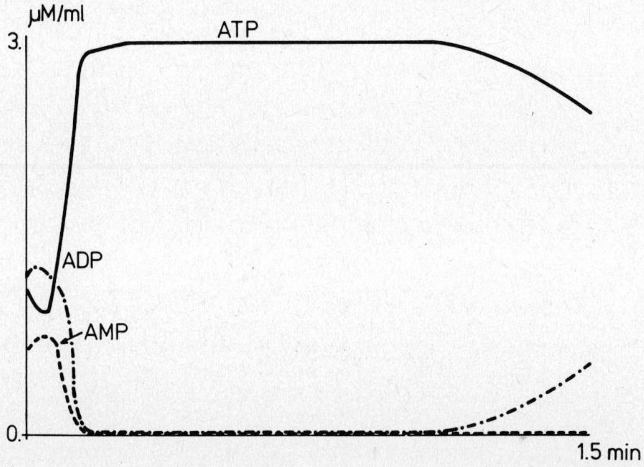

Fig. 3 Response of the adenylates in the model to the addition of glucose (aerobic conditions)

In the anaerobic case the initial drop of ATP is followed by the onset of oscillations, provided the ATP consumption rate lies in a certain range. These oscillations are charcterized by the phase relations between intermediates and cofactors as stated above. Fig. 4 shows the outcome of a typical experiment in yeast cells: ATP oscillates with a phase difference of 180° to ADP and AMP and with equal phase to the energy charge.

The same oscillatory behavior can be realized by the model system as is shown in Figs. 5a and 5b.

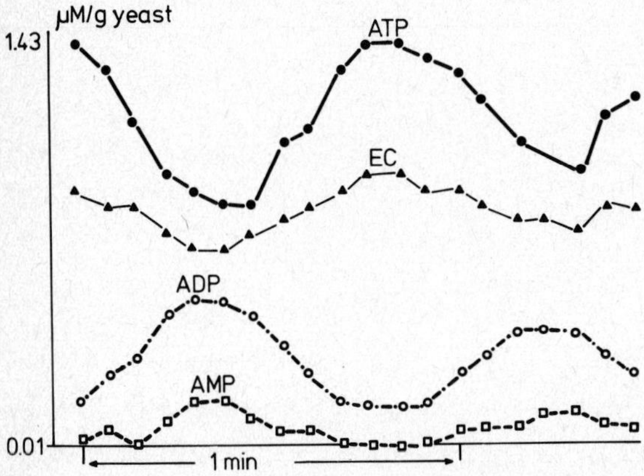

Fig. 4 Oscillations of the adenylates and the energy charge in an anaerobic suspension of yeast cells initiated by the addition of glucose

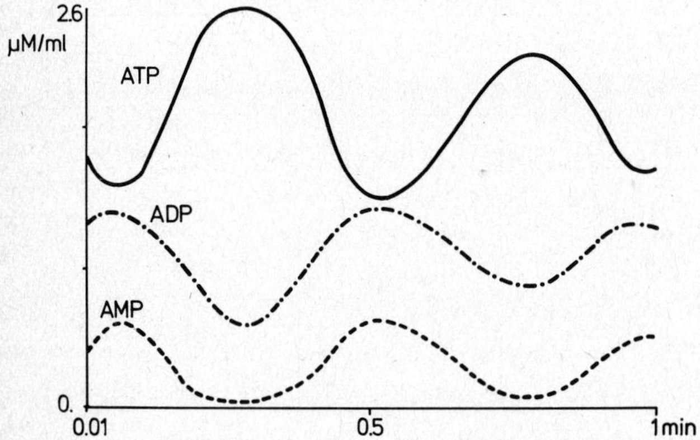

Fig. 5a Oscillations of the adenylates in the model

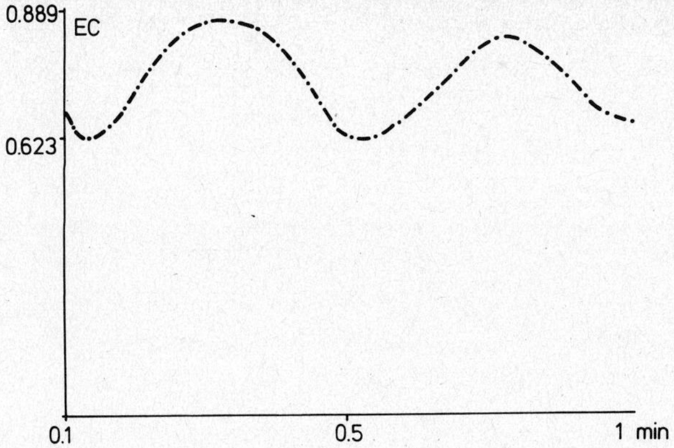

Fig. 5b Oscillation of the energy charge in the model

The comparison between model an experiment shows that even a simplified model, adequately designed, is able to simulate the dynamic behavior of multienzyme systems. So one can conclude that the subsystem of glycolysis chosen contains essential elements determining the dynamics of the whole system.

It should be noticed that unicellular organisms are best fitted for a mathematical description because the experiments are performed with a huge number of cells, so the interindividuel variabilitiy disappears.

In the next section an example is given, how the technique of simulation can be applied to medicine.

III Application to Medicine: Inhibiton of Xanthineoxidase by Allopurinol in Hyperuricemia Therapy

1. Some general Remarks on the Simulation of Metabolic Disorders

The simulation of biochemical pathways can be applied to diseases which are caused by metabolic disorder. Metabolic disorders like hyperuricemia or phenylketonuria are related to the malfunction of one or several regulatory enzymes. In the catabolism of purines for instance the total or partial loss of feedbackinhibition of the enzyme phosphoribosyl-pyrophosphate amido transferase by the purine nucleotides IMP, AMP and GMP causes an overproduction of uric acid which may lead to gouty attacks.

In general the interactions within metabolic networks of this kind are very complicated, so intuitive reasoning on cause and effect may lead to false predictions. A computer study based on a properly designed mathematical model of the pathway under study can help to get an insight into the dynamic and regulatory properties of the system before performing experiments or applying drugs. The aims of such a study can be defined as follows:

(i) analysis of the mechanism of a metabolic disorder caused by the malfunction of enzymes, primarily the effects of the malfunction of one key enzyme on the metabolite and activity pattern of the whole pathway.

(ii) analysis of the effects caused by the application of drugs which effect the activity of one or several enzymes on the whole pathway, that means investigating side effects. According to the amount and quality of the data two levels of applications can be distinguished:

a. The data are incomplete and assumptions habe to be made on intuitive grounds. In this case alternative hypothese on causes of disorders can be tested.

b. The data are sufficient for the conception of a realistic model. In this case the application of drugs can be simulated. If combined with pharmacokinetic compartimentmodels such a study can be of considerable practical value.

2. Biochemistry of the last steps in purine catabolism

In the last two steps in the catabolism of purines hypoxanthine is converted to xanthine which is converted to the end product uric acid. Both steps are mediated by one and the same enzyme xanthineoxidase E. C. 1.2.3.2 , so hypoxanthine is a competitive inhibitor of xanthine and vice versa. Allopurinol is a strong inhibitor of xanthineoxidase and is frequently used in the therapy of hyperuricemia, which is caused by some metabolic disorder in purine metabolism resulting in overproduction of uric acid. (gout). Besides the enzymatic degradation hypoxanthine can be eliminated from the cell by diffusion, so after the application of allopurinol an increased level of urinary oxypurines is observed.

3. The Model

The reaction scheme

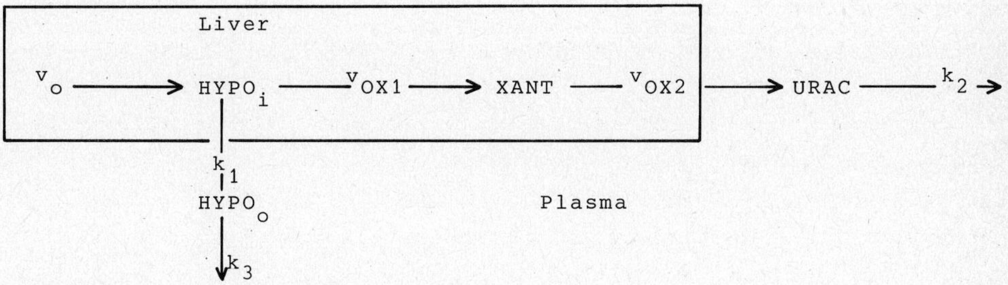

follows the equations

$$\dot{[HYPO]}_i = v_o - v_{OX1} - k_1 \, [HYPO]_i \tag{19}$$

$$\dot{[XANT]} = v_{OX1} - v_{OX2} \tag{20}$$

$$\dot{[URAC]} = \alpha v_{OX2} - k_2 \, [URAC] \tag{21}$$

$$\dot{[HYPO]}_o = \alpha k_1 \, [HYPO]_i - k_3 \, [HYPO]_o \tag{22}$$

where k_1 is an elimination constant for the transition liver plasma, and k_2 and k_3 are elimination constants for the elimination from the plasma. α is a scale factor which is needed to couple biochemical dimensions (μM/mg/day) to pharmacokinetic dimensions (mg%/day). α depends on the individuel compartiment volumes. The kinetic laws v_{OX1} and v_{OX2} are Michaelis-Menten laws with competitive inhibition by HYPO and XANT according to [14] and with assumed noncompetitive inhibiton by allopurinol.

$$v_{OX1} = V_{max} \frac{[HYPO] \; / \; (1 + [ALLO] / K_I)}{[HYPO] + K_{hypo}(1 + [XANT] / K_{xant})} \tag{23}$$

$$v_{OX2} = V_{max} \frac{[XANT] \; / \; (1 + [ALLO] / K_I)}{[XANT] + K_{xant}(1 + [HYPO] / K_{hypo})} \tag{24}$$

The subscripts i and o refer to concentrations within the cells (i) and in the plasma (o).

Allopurinol is applied during the time intervall (t_1, t_2) using a step function:

$$ALLO = \begin{cases} 0 & t \notin (t_1, t_2) \\ 2\,K_I & t \in (t_1, t_2) \end{cases} \qquad (25)$$

4. Results

Since this study is intended to give only qualitative results the constants are not fitted to individuel measurements. For a physiologically reasonable set of paramenters (cf. Table 2) the system shows the following response to the addition of allopurinol (cf. Figs. 6a and 6b). The concentrations of hypoxanthine and xanthine

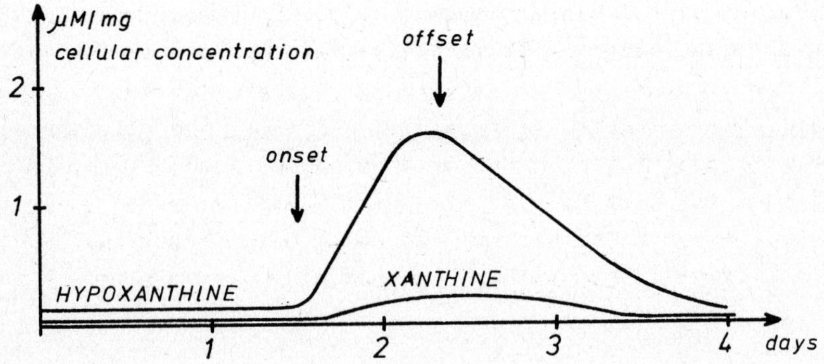

Fig. 6a Progress curves of cellular hypoxanthine and xanthine before and after the perturbation by allopurinol

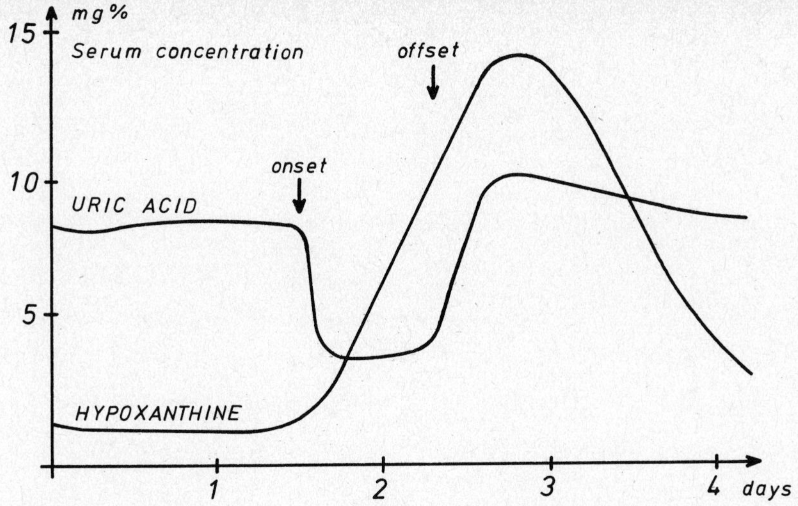

Fig. 6b Progress curves of serum uric acid and hypoxanthine before and after the application of allopurinol

rise within the cells, whereas the serum concentration of uric acid falls. The cellular level of hypoxanthine rises much higher than the level of xanthine because the two enzymatic steps are both mediated by the same enzyme. In addition to that the accumulation of hypoxanthine inhibts the conversion of xanthine. The level of serum oxypurine markedly increases. This is in qualitative agreement with measurements performed on gouty patients [15] . Because of accumulation of cellular hypoxanthine the offset of allopurinol treatment leads to an overshoot of serum uric acid concentrations. For a transient time, the level is higher than before treatment. If a critical value of uric acid concentration is reached, a gouty attack may occurr, e. d. the precipitation of sodium urate crystals. This may be a possible mechanism , which explains, why sometimes allopurinol therapy triggers a gouty attack in patients with high uric acid levels. So in the beginning of allopurinol therapy one should avoid a sudden offset of treatment as long as the primary overproduction of purines is not reduced.

5. Discussion

The application of the simulation technique to medicine is not limited by numerical or computational problems.

The main problem is a reasonable compilation of experimental data involving decisions wether data of in vitro studies are physiologically reasonable or not. In order to achieve results which are of practical value in therapy it is necessary that clinical biochemists, enzymologists and biomathematicians cooperate.

Appendix:Tables

Constant	Dimension	Value
v_o	μmoles/min ml	24.0
v_{max}(PFK)	μmoles/min ml	33.0
$v_{max}(V_R)$	μmoles/min ml	20.0
$k_{ATP}(V_{SINK})$	1/min	1.0
k_s	1/min μmole	6.0
K_{ATP}(PFK)	mM	0.05
K_{F6P}	mM	0.03
K_{AMP}	mM	0.01
k_1	1/min μmole	100.0
k_2	1/min μmole	50.0
L		250.0
K_m(ATP)	mM	0.01
K_m(FDP	mM	1.0
N_o	μmoles/ml	3.3

Table 1:
Constants used in the model of glycolysis

Constant	Dimension	Value
v_o	μ moles/mg/day	7.0
v_{max}	μ moles/mg/day	15.0
K_{hypo}	mM	0.025
K_{xant}	mM	0.004
k_1	1/day	0.8
k_2	1/day	22.0
k_3	1/day	3.0
α	mg%/μ moles/mg	28.0

Table 2:
Constants used in the model of hyperuricemia therapy

References

1 Otten, H.A., and Duysens, L., N., J. theor. Biol. 39, 387-396 (1973)

2 Reich, J. G., and Selkov, E. E., Biosystems 7,1,39-51 (1975)

3 Tikhanov, A. N., Mat. Sb. 22,64,193-201 (1948)

4 Walter, C. F., J. theor. Biol. 27,259-272 (1970)

5 Chance, B., Betz, A., and Hess, B., Biochem. Biophys. Res. Comm. 16,2 (1964)

6 Tornheim, K., and Loewenstein, J. M., J. Biol. Chem. 249, 3241-47 (1974)

7 Frenkel, R., Arch. Biochem. Biophys. 125,157-165 (1968)

8 Hess, B., Boiteux, A. and Krüger, J. in "Advances in Enzyme Regulation", Vol. 7, 149-167 (1969)

9 Hess., B., Ergebnisse der exp. Med. 9,66-87 (1971)

10 Selkov, E. E., Eur.J.Biochem. 4,79-86 (1973)

11 Goldbeter, A., FEBS Letters 43, 327-330 (1974)

12 IBM Scientific Subroutine Package, Version III, Subrocutine HPCG

13 Kopperschläger, G., Bähr, M.L. and Hofmann, E., Acta biol. med. german. 19,691-704 (1967)

14 Lalanne, M. and Willemot, J., Int. J. Biochem. 6,479-484 (1975)

15 Wyngaarden, J., in "Advances in Metabolic Disorders", Vol. II, 1-67 (1965)

SOME REMARKS ON THE PHYSICAL BASIS OF PHARMACOKINETICS

Rudolf Repges

Technische Hochschule Aachen

The starting point for the following lectures is a system of linear differential equations of the type

$$(1) \quad \frac{d\,(V_p\, c_{ps})}{dt} = \phi_{pqs}(t) + \sigma_{ps}(t) \qquad \begin{array}{l} 1 \leqq p,q \leqq a \\ 1 \leqq s \leqq b \end{array}$$

describing the history of the amount $V_p\, c_{ps}$ of a substance s in a compartment p, c_{ps} being its concentration and V_p the volume of this compartment, in terms of the flow ϕ_{pqs} of substance s from compartment p to compartment q and of the source or sink σ_{ps}, if the substance is produced or destroyed e.g. by chemical reactions within the compartment or at its boundary.

Equation (1) is nothing but a balance equation for each compartment, assuming however a rapid mixing or referring only to the total mass of the substance under consideration rather than to its concentration.

Equation (1) also applies to stochastic models; in this case ϕ corresponds to the migration term and σ to the birth and death rate.

The crucial part consists essentially in finding an appropriate expression for ϕ and σ. The usual formulations are:

(2) $\phi_{pqs}(t) = \lambda_{pqs}\, c_{ps}(t)$, if there is a path from compartment p to q for substance s
$= 0$, if there is no such transition.

(3) $\sigma_{ps}(t)$ $= \delta(t-t')$, if there is an injection of substance s into compartment p at time t'
$=$ constans, $t' < t < t''$, if there is an infusion during this interval
$= f(c_{p1}, c_{p2},\ldots,c_{pb})$, if there is a chemical reaction between c_{ps} and other substances within compartment

p, including receptors, as it is assumed in pharmacodynamics.

Let us now look, whether there is some physical or physiological justification of equation (2) describing the transition or flow of s from p to q.

Physically speaking the problem consists in determining the densities ρ_s and velocities v_s as functions of space and time for all substances $1 \leqq s \leqq b$ of interest. This is a problem of the field theory, and the starting point are again the balance equations, generally speaking, of the scalars mass ρ_s and energy ε, of the momentum vector $\rho_s v_s$ and of the electromagnetic field tensors B and H with elements

$$B_{ij} = \varepsilon_{ijk} B_k, \quad B_{4i} = - B_{i4} = E_i \quad \text{(vacuum) and}$$

$$H_{ij} = \varepsilon_{ijk} H_k, \quad H_{4i} = - H_{i4} = D_i \quad \text{(matter)}$$

$\varepsilon_{ijk} = 1$ for even permutations and -1 for uneven permutations of (1,2,3).

Let us first assume that there is

(A1) no electromagnetic field

(A2) no spatial or temporal variation of the temperature.

Then the balance equations reduce to the balance of mass, ρ_s, and momentum, $\rho_s v_s$, reading

(4) $$\partial_t \rho_s + \partial_j (\rho v_j)_s = \sigma_s$$

(5) $$\partial_t (\rho v_i)_s + \partial_j (\rho v_i v_j - t_{ij})_s = (m_i + \rho f_i)_s$$

where σ_s and $(\rho f_i)_s$, the source term and the external forces, are assumed to be given, and t_{ijs} and m_{is}, the flow and the production of the momentum of substance s, are functions of the ρ_r and v_r and their derivatives and are characteristic for the material under consideration. Once these functions are known, the system is fully described.

By summing over all substances s we arrive to the balance equation for the total mass and the total momentum

(6) $\partial_t \rho + \partial_i (\rho v_i) = 0$

(7) $\partial_t (\rho v_i) + \partial_j (\rho v_i v_j - t_{ij}) = \rho f_i$

if we define

(8)
$$\rho = \Sigma \rho_s, \qquad \rho v_i = \Sigma \rho_s v_{is}, \qquad \rho f_i = \Sigma \rho_s f_{is},$$
$$t_{ij} = \Sigma (t_{ijs} - \rho_s u_{is} u_{js}) \quad \text{with} \quad u_{is} = v_{is} - v_i$$

and use $\sum_s m_{is} = 0$ and $\sum_s \rho_s u_{is} = 0.$

The only possibility to derive equation (2) is now to impose suitable restrictions on the expressions for the t_{ij} and m_i.

The first restriction as usually done in fluid mechanics is

(A3) t_{ij} and m_i may depend only on ρ and v and their first derivatives $\partial_i \rho$ and $d_{ij} = \frac{1}{2} (\partial_i v_j + \partial_j v_i)$.

Simple arguments concerning the isotropy of fluids and the invariance against Euclidean transformations lead to the expression

(9) $t_{ij} = - (p(\rho) - \lambda(\rho)\, d_{ii})\, \delta_{ij} + 2\, \mu(\rho)\, d_{ij}$

Inserting (9) into equation (7) results in the famous Navier-Stokes-equation. But we will further restrict ourselves to fluids

(A4) being nonviscous and incompressible, i.e. $\lambda = \mu = 0$ and $\partial_t \rho = \partial_i \rho = 0$;

(A5) allowing to neglect all terms quadratic in the relative velocities;

(A6) having functions m_{is} linear in v_{ir} and $\partial_i \rho_r$.

These three assumptions allow us to write

(10) $t_{ij} = - p(\rho)\, \delta_{ij}$

(11) $t_{ijs} = - p_s(\rho_1, \rho_2, \ldots, \rho_b)\ \delta_{ij}$

(12) $m_{is} + \partial_j t_{ijs} = \sum_r g_{isr}\ v_{ir} + \sum_r h_{isr}\ \partial_i \rho_r$.

Let us now introduce this expression in (5). Some simple computations, using the barycentric derivation $d_t = \partial_t + v_j \partial_j$, and subtracting the last equation (s=b) from the the others to avoid linear dependence, yield - in a convenient matrix notation -

(13) $d_t v + w = Gv + H\partial\rho + \Delta f.$

Our final aim to find an expression for the flows $\rho_s v_{is}$ or the relative flows $\rho_s u_{is}$ seems to be possible, if we write

(14) $v = G^{-1}\ (d_t v + w - H\partial\rho - \Delta f)$ or, with $v = L\ u$:

(15) $u = L^{-1}\ G^{-1}\ (d_t Lu + w - H\partial\rho - \Delta f)$,

and dare our next anyhow severe simplification by neglecting

(A7) the divergence term w

(A8) the inertial term $d_t v$

(A9) the external forces Δf.

Under these assumptions it is possible to reduce our last equation to the well-known Fick-equation

(16) $u = - D\partial\rho$

with $D = L^{-1}\ G^{-1}\ H$. This is already similar to the pharmacokinetic set up, and if we now bravely neglect also all interactions between the substances by letting D be diagonal, e.g.

(A 10) $D_{sr}\ \rho_s = - \lambda_s \delta_{rs}$

and reduce the gradient $\partial_i \rho_s$ to a jump $\Delta_{pq} \rho_s$ between the compartments p and q as the only spatial dependence of ρ_s (A 11), we achieve our final result

(17) $(\rho u_i)_s = \lambda_s' \Delta_{pq}\rho_s$

(18) or $\phi_{pqs} = \lambda_{pqs} (c_{ps} - c_{qs})$.

Whether it is possible to further simplify this result to equation (2) - by an assumption A 12:

(A 12) $\lambda_{pqs} (c_{ps} - c_{qs}) = \lambda_{pqs} c_{ps}$

may be left as a matter of discussion.

Our aim has been to explicitely note all assumptions - A 1 unto A 12 - which are tacitly made in pharmacokinetics, such that one has the possibility to decide which of these assumptions he can justify in his specific problem and which of them might probably be an over-simplification.

References

Becker, E. and W. Bürger:
Kontinuumsmechanik.
Teubner (Stuttgart) 1975

Glansdorff, P. and J. Prigogine:
Structure, stability and fluctuations.
Wiley (New York) 1971

Müller, J.:
Thermodynamik.
Bertelsmann (Düsseldorf) 1973

Rescigno, A. and G. Segre:
Drug and Tracer Kinetics.
Blaisdell 1966

Mathematical models in the study of drug kinetics

Giorgio SEGRE - Institute of Pharmacology, University of Siena, Italy.

****

The theory of compartments represents the mathematical basis of the study of pharmacokinetics. The application of such a theory to the problems of drug kinetics, that is to the study of the movements followed by drug in the organism, in isolated organs or apparatuses, began with the paper of Gehlen (1), followed by those of Teorell (2) and of Beccari (3). After the second world war the theory was clearly stated and applied with its different derivations by several Authors: Dost (14), Rescigno and Segre (5), Atkins (6), Notari (7), Wagner (8), Jaquez (9), Gladtke (10). Symposia have been held on this topic (11,12) and scientific journals have been published on pharmacokinetics (J. of Pharmacokinetics and Biopharmaceutics, Clinical Pharmacokinetics), whereas a high proportion of the papers which appear in the journals of clinical pharmacology are dealing essentially with problems of pharmacokinetics. The health autorities in the different countries require a detailed pharmacokinetics analysis for accepting a new drug and most of the leading drug companies have developed laboratories of pharmacokinetics.

The theory of compartments is based on a mathematical model formed by a system of linear equations of first order with constant coefficients, where for each component i (called compartment), the following relationship is considered valid

$$dX_i/dt = \Sigma_j k_{ji} X_j - \Sigma_j k_{ij} X_i \qquad \text{(I)}$$

$(i = 1,.......,n; \; j = 1,......,n; \; i \neq j)$

In this model the components of the system of equations are homomorphic to the components of the real system under study, that is each component of the mathematical model corresponds to one or more components of the real system.

Given the initial conditions $X_i(0)$, the solution of the previous system is formed usually by a multiexponential equation

$$X_i(t) = \sum_r A_{ir} e^{-a_r t} \qquad (r=1,\dots,n) \qquad \text{(II)}$$

where one of the a_r's can be zero.

In certain cases one has in eq.(II) terms of the type

$$e^{-at},\ t\,e^{-at},\ t^2 e^{-at},\ \dots,\ t^{r-1} e^{-at} \qquad \text{(III)}$$

or terms of the type

$$t^b e^{-ct} \cos\lambda t\ ,\ t^b e^{-ct} \sin\lambda t \quad (b = 0,1,2,\dots) \qquad \text{(IV)}$$

Between the matrix of the terms k_{ij} in (I) and eq.(II) the following relationship can be established

$$|k| = |A|\ |a|\ |A|^{-1}$$

where $|a|$ is a diagonal matrix of the form

$$\begin{vmatrix} a_1 & & & & \\ & a_2 & & & \\ & & \cdot & & \\ & & & \cdot & \\ & & & & a_n \end{vmatrix}$$

In eq.(I) the variables $X_i(t)$ indicate the amounts or the concentrations in the different compartments i.

In the latter case the driving force of the system is given by the concentration difference between compartments and a diffusion through membranes, following Fick's law, is assumed. In the former case the driving force is given by the blood perfusion of the organs or of the tissues ; such a force is often assumed when an organ clears a drug from the circulation and excretes it in the excretion fluid (bile, urine) at a concentration higher than that of the blood.

The constants k_{ij} are called transfer constants. The compartments are defined as ideal volumes (or pools) in which each particle present in it has equal probability of exit; this corresponds to admit that the material present in a compartment is mixed at a rapid rate

in comparison with the transfer process to other pools.
Moreover a compartment may correspond to a localization, that is to a distribution volume, or to a chemical state; in effects the reagents and the reaction products in a given volume are represented by as many compartments which are coincident in the real space but not in the representative space.

The transfer constants which appear in eq. (I) have the dimension t^{-1}; a transfer constant k_{ij} represents the fraction of the material present in compartment i which enters compartment j in one unit of time.

Compartmental systems, in which the compartments are given by an organ or by a collection of organs and the transfer constants are related to the blood flowing rate of these organs, are well known in pharmacokinetics and provide often, as in the case of thiopental, results that are in good agreement with the experimental data (13).

The theory of compartments as applied to pharmacokinetics is a theory that assumes that the system is linear and non-discriminating. For instance if one suppose that the particles present in one compartment (like the erytrocytes) outflow from the system following a survival curve (that is the older red cells disappear sooner than the younger ones), one has a discriminating, linear compartmental system. Moreover the theory of compartments appears to be essentially a phenomenological one with a prevailing formal structure, which is useful to organize our knowledge in the field of drug kinetics and to establish certain relationships among the measurable quantities. On the other hand the theory possesses a noteworth heuristic value: the theory compels to a clear statement of the assumptions made and of the conditions of validity of the solutions, and permits to obtain all the information content of a given experiments, to express in few parameters an array of data (reduction of data), to stress the weak points of the assumptions, to suggest a better or a new experimental design or even new experiments.

In relation to compartmental analysis it is possible to give a more precise definition of the term "drug": a drug is a substance not belonging to the composition of the living system that influences partially the behaviour of the system; but a drug is also a material already present in the living system, provided it is introduced in amounts which perturb some of the steady states of the system.

Very often a drug is studied with respect to its effects on the transfer constants that characterize the kinetics of a given living component mostly in steady state, or the kinetics of another drug.

Moreover the theory of compartments can be applied to the system in steady state when tracers are used to investigate them; in this case eq. (I) is valid even when the reactions among the mother substances in steady state are not linear, provided the variables are represented by the activities or by the specific activities.

When the values of a given transfer constant show a bimodal distribution in a given population, such a transfer constant can be used as a genetic marker, as may happen for the plasma disappearance rate of isoniazid (fast and slow acetylators); in this way the doors of pharmacogenetics were opened.

Applications of compartmental models can be found in ecology and in epidemology as well (see Watt (14)).

When the system is no more time-determined but some of the transfer constants change with time, we are in the field of enzyme induction or in that of the discrimination between normal and pathologic cases or in the position of using such data for prognostic purposes, or even in the field of circadian rythms. When a system contains transfer constants with negative values, we are in the field of the negative feedback, and in the chapter of the control, like the hormonal control. Therefore compartmental analysis is also connected with system analysis and appears as an applied mathematical theory with imposing simplicity; this theory appears to be a tool with many applications in Biology and Medicine, and represents probably the most effective example of application of Mathematics to Biology.

It is particularly useful, in dealing with compartmental models

and with their mathematical models, to employ the Laplace transforms. This transformation belongs to linear algebra; in this case the multiplication corresponds to the integration and to the convolution and the division corresponds to the deconvolution.
The Laplace transform of eq. (I) is given by eq. (V)

$$s\, x_i(s) - x_i\,(0) = \Sigma_j\, k_{ji}\, x_j(s) - K_i\, x_i(s) \qquad (V)$$

where $K_i = \Sigma_j\, k_{ij}$. Eq. (V) can be written

$$(s + K_i)\, x_i - \Sigma_i\, k_{ji}\, x_j = x_i(0) \qquad (VI)$$

System (VI) corresponds to an algebric equation in s; by Kramer's rule one obtains the solution for x_i. Let us assume $x_1(0) = x_0$ and $x_i(0) = 0$, then

$$x_i(s)/x_0 = (-1)^{i+1}\, D_{1,i}/D \qquad (VII)$$

where

$$D = \begin{vmatrix} (s + K_1) & - k_{21} & \cdots - k_{n1} \\ - k_{12} & (s + K_2) & \cdots - k_{n2} \\ \cdots & \cdots & \cdots \\ - k_{1n} & - k_{2n} & \cdots (s + K_n) \end{vmatrix}$$

and $D_{1,i}$ is the minor D obtained by suppressing the 1-st row and the i-th column.

The solution

$$x_i(s)/x_0 = p(s)/q(s) \qquad (VIII)$$

is given by a fraction in which the roots of $D = q(s) = 0$ are the poles and correspond to the eigenvalues of D, whereas the roots of $D_{1,i} = p(s) = 0$ are the zeros of the function. Both p(s) and q(s) are polynomials in s, with q(s) of lower degree than p(s).
Partial fraction expansion of (VIII), when the function (VIII) has

n poles, gives

$$\frac{x_i}{x_0} = \sum_r \frac{A_r}{s + a_r} \qquad \left(r = 1,\ldots,n\right) \qquad \text{(IX)}$$

where $A_r = \left.\frac{(-1)^{i+1} D_{1,i}}{D'}\right|_{s = -a_r}$

and $D' = \frac{dD}{ds}$.

The antitransform of (IX) corresponds to the multiexponential function (II), and when we have a root equal to zero or multiple roots, or complex conjugate roots, one term of (II) is a constant or we have the solutions shown by eq. (III) and (IV).

When the material is not introduced into the system as an impulse formed by the dose x_0, but following a function G(t), the previous equation (VIII) $x_i(s) = x_0\, p(s)/q(s) = x_0\, f(s)$

becomes $x_i(s) = g(s) \cdot f(s)$,

which corresponds to the convolution between the two function G(t) and F(t), i.e.

$$X_i(t) = G(t) * F(t) = \int_0^t G(\tau - t)\, F(\tau)\, d\tau \ .$$

By using the Laplace transformation it is possible to represent the mathematical model of a system of compartments by an oriented graph formed by nodes and oriented arms, on which it is possible to carry out the operations that lead to the solution of the eq. (VI). In this case one can define as a transfer function between j and i the ratio

$$x_j(s)/x_i(s) \qquad \text{(X)}$$

between the two functions, which corresponds to the deconvolution between the functions in the time domain. The antitransform of eq.

(X) can be called the weighting function between j and i, and corresponds to the kinetics that can be observed in j when the material is introduced in i as a unit impulse. It can be recalled by remembering eq.(VIII) that

$$x_j/x_i = (-1)^{(i+j)} D_{1,j}/D_{1,i} \quad .$$

When i and j are adjacent, the previous ratio becomes

$$\frac{x_j}{x_i} = \frac{k_{ij}}{s + K_j} = T_{ij}$$

whose antitransform is equal to $k_{ij}\ e^{-K_j t}$.

Therefore in the graph which represents a compartmental system, a node represents the Laplace transform of the function which describes the kinetics of the material in a given compartment, and the arm T_{ij} represents the transfer function between two adjacent nodes. By using the Mason's rules it is possible to calculate the transfer function between any two nodes of a system.

The rules are the following:

1) the material introduced into comp. 1 is represented by a node (the node zero) connected with comp. 1 by an arm whose value is equal to $\frac{1}{s + K_1}$
2) two arms in series can be substituted by an arm equal to their product
3) two arms in parallel can be substituted by an arm equal to their sum
4) a recycling formed by two o more arms i(1,2,....,n) that initiates and terminates in the same mode can be eliminated by multiplying every arm entering a node belonging to the recycling by the function $1/(1-T_{ii})$, where T_{ii} is formed following the previous rules, and is of the type

$$T_{ii} = \sum_r \prod \frac{k_{ir}}{s + K_i} \quad (r = i,j\ldots\ldots)$$

Another more straightforward method of solving the graph, which avoids the difficulties of applying the Mason's rules when multiple recyclings are present, is the following:
let us write eq. (V) as

$$x_i = \sum_j \frac{k_{ji}}{s + K_i} x_j + \frac{x_0}{s + K_i} = \sum_j T_{ji} x_j + T_{0i} x_0 \qquad \text{(XI)}.$$

It is possible introduce the following definitions:
path = a succession of arms between adjacent nodes;
elementary path = a path which does not passes twice through the same node;
elementary cycle = a path starting from a node and ending in it without passing twice through the same node;
strong graph = a graph in which from each node it is possible to reach any other node.

Eq. (XI) can be written in extenso:

$$\begin{cases} x_1 \quad -T_{21} x_2 \cdots\cdots -T_{n1} x_n = T_{01} x_0 \\ -T_{12} x_1 + x_2 \quad \cdots\cdots -T_{n2} x_n = T_{02} x_0 \\ \cdots\cdots\cdots\cdots\cdots\cdots \\ -T_{1n} x_1 - T_{2n} x_2 \cdots\cdots + x_n = T_{0n} x_0 \end{cases} \qquad \text{(XII)}$$

If we suppose $T_{01} \neq 0$, $T_{0i} = 0$ $(i = 2,\dots,n)$, without loss of generality, because of the superposition principle, we can solve the system and obtain

$$x_i/x_0 = \sum \prod M_{ip}/M_p \qquad \text{(XIII)}$$

Eq. (XIII) says that the solution of (XI) and (XII) is given by the sum of the products of the solutions of the strong graphs which are connected in such a way that one is the input of the other.

For each strong graph the term M_{ip}/M_p is given by a fraction formed by a numerator and a denominator:

The denominator is equal to 1 minus the value of each elementary cycle, plus the products of the values of the elementary cycles which do not possess common nodes, taken 2 by 2, 3 by 3 and so on, with sign + or - depending on whether the number of the cycles is even or odd.

The numerator is formed by the sum of the values of the elementary paths going from node 0 to node i, minus the sum of each of the elementary paths that are not touched by said elementary paths, plus the products of the values of elementary cycles not touched by the said elementary paths and without common nodes, taken 2 by 2, 3 by 3, and so on, with sign + or -, depending on whether the number of the cycles is even or odd. (15,16).

In the same way it is possible to compute the transfer function between any two compartments of the system , i and j. This function is equal to the ratio of the previously seen transfer functions, calculated for i and for j. If i and j belong to the same strong graph, the denominator of eq. (XIII) disappears and we have the ratio of the numerators computed as said before. This corresponds to the results of an experiment in which the dose has been introduced in a given compartment (connected with comp. i) or in comp. i itself and the measurement has been carried out in i and in j.

A different value for the ratio x_j/x_i is obtained when the dose is introduced in one experiment in comp. i and in another experiment in comp. j, as sometimes is done in determining the structure of the graph which connects the two compartments.

The rules for the graph formation can be extended to include also the zero-order reactions (17).

One of the advantages of the use the Laplace transforms in compartmental analysis in due to the fact that the convolution and the inverse operation, that is the deconvolution, become a multiplication and a division.

Such operations, particularly interesting in model building, can be approximated as follows: if $Y(t) = G(t) * X(t)$ is the convolution of $G(t)$ and $X(t)$, and if $X(t)$ is approximated by the areas subtended by the curve in each equal interval time $\Delta\theta$, that is by $X_k\Delta\theta$

(k = 0,1,2,.....n), and G(t) is represented by the values of the function at the corresponding time intervals, G_0, G_1, G_2......, then one obtains:

$$Y_n \Delta\theta = \sum_{k=0}^{n} G_{n-k} X_k \Delta\theta \qquad \text{(XIV)}$$

when $Y_n \Delta\theta$ are the approximated values of the areas subtended by the curve Y(t).

By supposing for simplicity $\Delta\theta = 1$, one can write eq. (XIV) as the matricial equation (XV)

$$\{Y_n\} = |G| \{X_k\}$$

with $\{Y_n\}$ and $\{X_k\}$ indicating column matrices and

$$|G| = \begin{vmatrix} G_0 & 0 & 0 & \ldots\ldots 0 \\ G_1 & G_0 & 0 & \ldots\ldots 0 \\ G_2 & G_1 & G_0 & \ldots\ldots 0 \\ \ldots & \ldots & \ldots & \ldots \\ G_n & G_{n-1} & G_{n-2} & \ldots\ldots G_0 \end{vmatrix} \qquad \text{(XV)}$$

is a lower triangular matrix.

If $|G|$ and $\{X_k\}$ are known, it is possible to calculate $\{Y_n\}$. If however $\{Y_n\}$ and $\{X_k\}$ are known, it is possible to calculate $\{G_k\}$ by the equation (XVI):

$$\{G_k\} = |X|^{-1} \{Y_k\} \qquad \text{(XVI)}$$

where

$$|X|^{-1} = \begin{vmatrix} X_0 & 0 & 0 \ldots & 0 \\ X_1 & X_0 & 0 \ldots & 0 \\ \ldots & \ldots & \ldots & \ldots \\ X_n & X_{n-1} & X_{n-2} \ldots & X_0 \end{vmatrix}$$

A better approximation is given by applying a trapezoid approximation for the areas.

This method has the drawback that Y_n, or, in the inverse process, G_k, tend to oscillates as n or k increase due to the propagation of the errors.

The analysis of the weighting function $G(t) = L^{-1}\ g(s)$ in the neighbourhood of $t = 0$ is of great interest in relation to the lenght of a path, that is to the number of arms which connect any two compartments i and j such that $x_j/x_i = g(s)$.

Since $\lim_{s\to\infty} g(s) = 0$, g(s) is an infinitesimal for $s\to\infty$ and the behavior of G(t) at $t\to 0$ is related to that of s g(s) as $s\to\infty$.

We define the precursor order of a transfer function g(s) the value r for which

$$\lim_{s\to\infty} s^r\, g(0) = M \quad \left(\begin{matrix}\neq 0\\ \neq \infty\end{matrix}\right)$$

This limit M is called the "precursor principal term".

In this case, from a known property of the Laplace transforms,

$$G(0) = G'(0) = \ldots\ldots\ldots\ G^{(r-2)}(0) = 0$$
$$G^{(r-1)}(0) = M$$

where (18)

$$G^{(i)}(0) = \left.\frac{d^i\, G(t)}{dt^i}\right|_{t=0}$$

The value of r gives the number of arms connecting i to j.

It can be recalled that if g(s) is the transfer function between comp. 0 and comp. j, that is

$$g(s) = x_j/x_o$$

then r gives the number of arms connecting comp. 0 to comp. j, and $G^{(r-1)}\ (0) = X_j^{(r-1)}\ (0)$.

It can moreover be shown that if $X_j^{(p)}(0) = 0 \quad (p< m)$, $X_i^{(m)}\ (0) \neq 0$, and $G^{(q)}(0) = (0) \quad (q< r-1)$, then

$$G^{(r-1)}(0) = C \lim_{t \to 0} \frac{X_j(t)}{t^r X_i(t)} \qquad \text{(XVII)}$$

where $C = \frac{(m + r)!}{m!}$

It can also be observed that the order of the precursor corresponds to the difference of the degree in s between numerator and denominator of the transfer function $g(s) = p(s)/q(s)$.

The evaluation of r through eq. (XVII) is very useful in model building. Even if the evaluation of the limit becomes more difficult as r increases, the determination of the limit for r = 1 gives a clue to assess whether two compartment are adjacent or not.

Other criteria to evaluate the goodness of a given model, that is whether a given model is compatible with the experimental data, is given by the criterium given by Hearon (19)

$$T . X_2(T) \leqslant Q/e$$

for an irreversible catenary system of 2 compartments, where $Q = \int_0^\infty X_2(t)\, dt$ and T is the time of the maximum of the curve X_2.

If the inequality (area of the rectangle TX(T) is less than the area of the curve divided by e) is not valid, the model ought to be rejected.

The same formula is valid also for a system of 2 open compartments reversibly connected, with a unit dose, and is therefore valid for a precursor of 1-st order.

For a catenary system of n compartments the following inequalities ought to be followed

$$\frac{1}{e\, T_2} \geqslant \frac{X_2\,(T_2)}{Q_2} \geqslant \cdots\cdots\cdots \geqslant \frac{X_n(T_n)}{Q_n}$$

where the indexes refer to the compartments.

For a precursor of n-th order the following formula is valid

$$\frac{T_n X_n(T_n)}{Q_n} \leqslant \frac{(n-1)^n}{(n-1)!} e^{-(n-1)}$$

with the equality valid when in the catenary system all the transfer constants are equal (20).

The following property of the Laplace transforms is also useful in compartmental analysis

$$f(s)\Big|_{s\,=\,0} = \int_0^\infty F(t)dt.$$

In this way it is possible to estimate the areas of the curves obtained in experiments of pharmacokinetics; one has

$$x_i/x_o\Big|_{s=o} = (-1)^{i+1} \quad D_{1,i}/D\Big|_{s=o}$$

or, by using eq. (XIII),

$$x_i/x_o\Big|_{s=o} = \sum\prod M_{ip}/M\Big|_{s=o} \quad .$$

The analysis of the areas subtended by a curve is of particular interest for instance in the study of the bioavailability of a drug. In effects it is clear that is a given amount of material (dose) enters comp. i and this compartment is connected with outside by an arm equal to k_{io}, one has:

$$\text{amount excreted} = k_{io} \int_0^\infty X_i(t)dt$$

where $X_i(t)$ is the amount of the material and the time present in comp. i ($X_i(t) = V_i\ x_i(t)$, where V_i = volume of i, x_i = concentration).

Therefore if all the dose is excreted from i one has

$$\frac{\text{Dose}}{\int_0^\infty X_i\,dt} = k_{io};\ \frac{\text{Dose}}{\int_0^\infty x_i\,dt} = V_i\ k_{io} = \text{Clearance of i.}$$

If only a fraction of the dose (f·dose), reaches comp. i by an irreversible route, one has

$$\frac{\text{f·dose}}{\text{area}} = k_{io}\ .$$

This is the base to compare whether two preparations of a given drug are equally absorbed by intestinal route or have equal bioavailability.

A more complete analysis of intestinal absorption kinetics is given by the intestinal transfer function.

The kinetics in blood after i.v. injection of the dose x_o can be expressed as $x_o^i\ x_i(s) = x_i^i$ (where the superscript refers to the compartment where the dose has been introduced) and the kinetics in blood after oral administration of the same dose can be expressed as $x_o^1\ g(s)\ x_i(s) = x_i^1$. The precursor order of G(t) can be computed from the two curves (i.v. and oral), and g(s) is equal to

$$\frac{x_i^1}{x_i^i} \cdot \frac{T_{oi}}{T_{o1}} \quad .$$

Similarly the transfer function for the various bodily functions (f.i. renal excretion, biliary excretion) can be computed by the analysis of blood kinetics of a drug and its simultaneously determined kinetics in the excretion fluid (urine, bile).

An evaluation which is seldom carried out in pharmacokinetics, but which deserves attention, is the determination of the time spent by a drug in a compartmental system.

The following definitions are deducible from the properties of the compartments (21, 22):

Mean transit time $(\bar{t}_i)$ = the average interval of time spent by a particle from it entry to comp. i to its next exit,

$$\bar{t}_i = 1/K_i, \text{ and by recalling that } K_i = -\dot{x}_i/x_i \Big|_{t=o}$$

$$\bar{t}_i = -x_i/\dot{x}_i \Big|_{t=o}$$

where K_i = sum of all the ouflows from i.

This corresponds to the turnover time in tracer kinetics.

Mean total residence time $(\bar{T}_1)$ = the average interval of time spent by a particle in a compartment in all its passages though it

$$\bar{T}_i = \frac{1}{x_o} \int_o^\infty x_i(\tau)\, d\tau \,,\quad \bar{T}_i = \frac{x_i(s)}{x_o}\Bigg|_{s=o} = \frac{1}{k_{io}}$$

where x_i = concentration and x_o = initial concentration, or x = amount, x_o = dose, and k_{io} is the fractional rate at which the material leaves irreversibly comp. i.

If all the material is lost irreversibly from comp. i, then $\bar{T}_i = \bar{t}_1$.

It can be recalled that $\bar{T}_i = \frac{\text{area}}{\text{dose}} = \frac{-\Sigma A_i a_i}{\Sigma A_i}$

If a fraction r of the material leaving the compartment returns in it, another fraction r^2 will return one more time, and so on;

therefore $\bar{T}_i = \bar{t}_1 + r\bar{t}_i + r^2\,\bar{t}_i + \ldots\ldots\ldots\ldots = \bar{t}_i/(1-r)$;

(1-r) is the fraction of the material leaving the compartment that does not return to it

$$(1-r) = \phi_i = \bar{t}_i/\bar{T}_i = k_{io}/K_i$$

Residence time $R_i = \frac{\int_o^\infty x(\tau)d\tau}{\int_o^\infty x(\tau)d\tau} = \frac{-dx(s)/ds}{x(s)}\Bigg|_{s=o}$

R_i corresponds to the expected time a particle of the material will leave the compartment irreversibly; that is the expected time that a particle introduced at time t = 0 spends in the compartment and in other parts of the system before leaving comp. i for the last time.

It is also possible to compute the number of cycles followed by a particle, that is the average number of times a particle passes throughout i.

Be N_i the number of times a particle returns to the compartment i after its first passage through it. If there is no recycling, then $N_i = 0$; otherwise $N_i + 1$ will be the number of cycles.

The mean number of cycles is given by $\bar{T}_i/\bar{t}_i$ and therefore $N_i = \bar{T}_i/\bar{t}_i - 1 = 1/\phi_i - 1 = \kappa_i/k_{io} - 1$.

<u>Mean recycling time</u> $\bar{t}_{ii}$ = average interval of time spent by a particle from its exit from the compartment to its next reentry. The interval of time spent outside the compartment before being lost irreversibly is equal to $R_i - \bar{T}_i$. Therefore the average time interval spent outside comp. i during a single cycle is $\frac{R_i - \bar{T}_i}{N_i}$.

Therefore $\bar{t}_{ii} = \frac{R_i - \bar{T}_i}{N_i}$.

When we analyze two compartments (not necessary adjacent) i and j, and i is the precursor and j the unique successor (that is all the material entering j comes from i, as is usual in drug kinetics), by recalling that

$$\frac{1}{x_o}\int_0^\infty x_i(\tau)\,d\tau = \frac{1}{k_{io}} \quad \text{and} \quad \frac{1}{x_o}\int_0^\infty x_j(\tau)\,d\tau = \frac{1}{k_{jo}}$$

is evident that $\int_0^\infty x_j(\tau)\,d\tau$ is a fixed fraction of $\int_0^\infty x_i(\tau)\,d\tau$.

Let us call W_{ij} the <u>weight of transport</u> from i to j, equal to

$$\int_0^\infty x_j(\tau)\, d\tau \Big/ \int_0^\infty x_i(\tau)\, d\tau .$$

If, moreover

$$U_i = \int_0^\infty \tau\, x_i(\tau)\, d\tau \Big/ \int_0^\infty x_i(\tau) d\tau \text{ and } U_j = \int_0^\infty \tau x_j\, d\tau \Big/ \int_0^\infty x_j\, d\tau$$

one has $U_j - U_i = Z_{ij}$, the <u>transfer time</u> from i to j.

If i anf j are adjacent, Z_{ij} corresponds to the <u>mean transit time</u> through j. (see also (5), p. 163).

In general, if g(s) is the transfer function between i and j, that is $g(s) = x_j/x_i$, one can calculate

$$Z_{ij} = \frac{-dg/ds}{g}\Bigg|_{s=0} \quad \text{and} \quad W_{ij} = g(s)\Bigg|_{s=0}$$

that is the transfer time and the weight of transport from i to j.

It is easy to deduce the rules for computing such parameters in a compartmental system.

a) for connected systems: $g_{il} = g_{ij}\, g_{jl}$;

one has $W_{il} = W_{ij}\, W_{jl}$

and $Z_{il} = Z_{ij} + Z_{jl}$

that is the transfer times add and the weights multiply;

b) for systems in series: $g_{ij} = g_{ij'} + g_{ij''}$;

one has $W_{ij} = W_{ij'} + W_{ij''}$

and $W_{ij}\, Z_{ij} = W_{ij'}\, Z_{ij'} + W_{ij''}\, Z_{ij''}$

(weights add and weighted transfer times add)

c) for a system followed by a recycling: $g_{ij} = g_{ij'} \frac{1}{1-g_{jj}}$;

one has $W_{ij} = W_{ij'} \frac{1}{1 - W_{jj}}$

and $Z_{ij} = Z_{ij} + \frac{W_{ij}}{1 - W_{ij}} Z_{jj}$.

These rules can be applied like the Mason's rules to compute the transfer times between the compartments in a system.

The interest of these considerations becomes apparent when one considers that the effect of a drug is in most cases determined by its permanence in a given compartment and that the transfer among various compartments bring about a delay in the appearance of a drug in a given compartment and of its effect.

As seen before, $x_i(s)/x_o$ is the transfer function of a drug from the compartment in which has been introduced as an impulse of value x_o at time $t = 0$ and comp. i. If the introduction follows the function $G(t)$, one will have as the Laplace transform of the kinetics in compartment i, the function $g(s).x_i(s)$.

When the drug is infused at a rate k_o one has

$$\frac{k_o}{s} x_i(s).$$

If the infusion stops at time $t = t_1$, one has

$$\frac{k_o(1 - e^{-t_1 s})}{s} x_i(s)$$

This function corresponds to the difference of the two integral curves for $X_i(t)$, shifted by t_1.
Of particular interest is a sinusoidal infusion, in this case the response will be given by a sinusoidal function, whose amplitude and phase change with the frequency of the imput.

The Bode diagrams of the logarithm of the amplitude ratio between the input and the output vs. the frequency of the input, and of the phase shift vs. the frequency provide informations on the compartmental structure of the system analogous to the informations obtained by compartmental analysis. It can be observed however that practically the circulatory delay determines the phase shift diagram.

If the dose x_o is repeated n times at equal time intervals, one has : $x_o\, x_i(s) \left[1 + e^{\tau s} + \dots + e^{n\tau s} \right]$, which, for large values of n, gives

$$\frac{x_o\, x_i(s)}{1 - e^{\tau s}} .$$

In the time domain one has, after the n-th administration,

$$X_i\,(t) = x_o \sum_i \frac{A_i\, e^{-a_i t}}{1 - e^{-a_i}}$$

This formula can be used to predict the drug levels after repeated administrations and is the base of the dosage regimen.

In connection with the case of the infusion, the problem arises of getting the final drug concentration in the blood

$$k_o X_i(\infty) = k_o \sum_i \frac{A_i}{a_i} \qquad \text{(XVIII)}$$

as soon as possible.

An empirical approach is given by injecting a given dose as an impulse together with an infusion at rate k_o (24).

In this case the concentration in compartment i follows a curve which shows a minimum after a given time.

Another approach is that of carrying out two consecutive infusions at different rates, the first at higher rate and maintained until values near the desired level are obtained (25).

Another aspect of compartmental analysis, which has acquired in these last years a noteworth importance, is the possibility of using also the kinetics of the pharmacologic effects after the necessary linearization (23). In this way it is possible to build up compartmental models which include the drug kinetics as well as the kinetics of the effects (23,26).

If also the kinetics of the metabolites is included, the system becomes very easily a large system, whose solution requires a careful analysis of its identifiability, the collection of data from several compartments, the subdivision of the system into subsystem, the analysis of the precursor-product relationships among them, and the simultaneous fitting of all the pieces of information to the same compartmental model, thus requiring the use of digital computers .

REFERENCES

1) GEHLEN W., Arch. Exp. Pathol. Pharmakol., 171, 541, 1933.
2) TEORELL T., Arch. int. Pharmacodyn., 57, 205, 1937, Ibidem, 57, 226, 1937.
3) BECCARI E., Arch. int. Pharmacodyn., 58, 437, 1938.
4) DOST F.H., Grundlagen der Pharmakokinetik, Thieme, Stuttgard, 1968.
5) RESCIGNO A., SEGRE G., Drug and Tracer Kinetics, Blaisdell, Waltham, 1966.
6) ATKINS G.L.,Multicompartment Models for Biological Systems, Methuen,London,1969.
7) NOTARI R.E., Biopharmaceutics and Pharmacokinetics - An Introduction, Dekker, New York, 1971.
8) WAGNER J.C., Biopharmaceutics and relevant Pharmacokinetics, Drug Intelligence Publ., Hamilton, Ill.,1971.
9) JACQUEZ J.R., Compartmental Analysis in biology and Medicine, Elsevier, Amsterdam, 1972.
10) GLADTKE E., VON HATTINGBERG H.M., Pharmacokinetik, Springer, Berlin, 1973
11) RASPE' G., Ed., Schering Workshop on Pharmackinetics, Pergamon, Oxford 1970.
12) TEORELL T., DEDRICK R.L., CONDLIFFE P.G., Pharmacology and Pharmacokinetics, Plenum Press, New York, 1974.
13) BISCHOFF K.B., DEDRICK R.L., J. Pharm. Sci., 57, 1346, 1968.
14) WATT K.E.F., System Analysis in Ecology, Ac. Press, New York, 1966.
15) RESCIGNO A., SEGRE G., Bull. Math. Biophys., 26, 31, 1964.
16) RESCIGNO A., SEGRE G., Bull. Math. Biophys., 27, 315, 1965.
17) GIORGI G., SEGRE G., in (11), pag. 89.
18) see (5) p. 180.
19) HEARON J.Z., Biophys. J., 1, 581, 1961.
20) LONDON W.P., HEARON J.Z., Mathem. Biosci., 14, 281, 1972.
21) RESCIGNO A., J. Theoret. Biol., 39, 9, 1973.
22) RESCIGNO A., GURPIDE E., J. Clinical Endocrinol. and Metabolism, 36, 263, 1973.

23) SEGRE G.,Il Farmaco,Ed.Sci.,23,907,1968.

24) MITENKO P.A.,OGILVIE R.I.,Clin.Pharmacol.Ther.,14,329,1973.

25) WAGNER J.C.,Clin.Pharmacol.Ther.,16,691,1974.

26) LEVY G.,GIBALDI M.,Ann.Rev.Pharmacol.,12,85,1972.

On some Applications of the Eigenvector Decomposition Principle in Pharmacokinetic Analysis

W. Müller-Schauenburg
Dep. of Nuclear Medicine
Medical Radiol. Institut
Röntgenweg 11
D-7400 Tübingen

Introduction

Pharmacokinetic analysis, as it is practised everywhere, is mainly a linear analysis /1/. Despite of wellknown nonlinear effects /2/ as protein-binding or active transport processes with their saturation phenomena, most authors apply models where all concentration versus time curves obey the superposition principle: i.e. all parts of a drug administered are assumed to travel independent of each other within the body.

Within this well established framework of linear models we confine to the well established field of compartmental models. We especially deal with situations where concentration versus time curves from blood or plasma form the basis of evaluations and predctions. The approach of the so-called eigenvector decomposition principle /3/ especially suites this basis.

We start from blood concentration data after an intravenous injection and assume that this concentration versus time curve is decomposed into exponential terms /cf. 4 (p. 146-16o), 5 (mainly p. 115-126), 6, 7/ by some least square fit for instance. We do not discuss the stability of such exponential fits. We simply offer a strategy which somehow reduces the complexity of multi-exponential calculations to the simplicity of the mono-exponential situation of the single compartment approach. The strength of our approach is that quite a lot of conclusions may be drawn without any further assumption on the special compartmental model and without a full determination of its parameters.

Among possible additional assumptions there is one which plays a central part in applying the eigenvector decomposition principle: the assumption of central elimination. It states that all substance is eliminated finally only from the central compartment (cf. below). Thus all lateral compartments are assumed to contribute only to the distribution of the drug but not to its metabolism or excretion.

Table I gives a survey on the applications of the eigenvector decomposition presented here and it shows which applications need the additional assumption of central elimination and which do not.

Application	confined to central elimination	aim
repeated dose	no	prediction of concentration curves only within the central compartment
	yes	prediction of the total drug amount in the system considered
volume of distribution	yes	to get the volume of distribution by calculating the infusion equilibrium from non-equilibrium i.v.-injection data
absorption analysis	no	reconstrucion of the fraction of dose absorbed as a function of time from central compartment data during absorption and after i.v.-injection

Tab. I

Mathematically the compartmental approach is a system of ordinary, linear differential equations with constant coefficients

$$\frac{d}{dt} m_j = \sum_{j'=1}^{n} k_{j'j} \, m_{j'} \qquad j=1....n \tag{1}$$

m_j : drug amount in the compartment j

$k_{j'j}$: transport constant $k_{j'\to j}$ from compartment j' to j

$k_{jj} := -k_{jO} - \sum_{j'=1, j'\neq j}^{n} k_{jj'}$ negative sum of all constants for transport leaving compartment j, where k_{jO} describes a final elimination from the system via compartment j

In matrix notation /8/

$$\frac{d}{dt} \underline{m} = \underline{\underline{k}} * \underline{m} \tag{2}$$

One of the variables m_1 is the drug amount in the central compartment. This compartment is the blood or plasma volume and some extra-vascular volume which is rapidly accessible to such a degree that it contributes to lowering the initial blood or plasma concentration value, which we get by extrapolating i.v. injection data back to injection time t = O.

For sake of simplicity we exclude those compartments which do not have a direct or indirect flow back to the central compartment. We do not get any information about those volumes from central compartment data and therefore we discard them.

What is the eigenvector decomposition principle ?

Eigenvector decomposition as a mathematical tool

System (1) od (2) may be solved either by reducing it to one differential equation of order n /cf. 9/ or by looking for special solutions of type

$$\underline{m}(t) = \underline{x} \cdot \exp(-\gamma \cdot t) \tag{3}$$

the so-called eigenvectors[1]. (For the approach by Laplace-Transformation see /1o/).

If we insert (3) into (2) we get the algebraic (time independent) matrix equation

$$-\gamma \, \underline{x} = \underline{\underline{k}} * \underline{x} \tag{4}$$

or ($\underline{\underline{E}}$ = unity matrix)

$$O = \left\{ \underline{\underline{k}} + \gamma \cdot \underline{\underline{E}} \right\} \underline{x} \tag{5}$$

Equating the determinant of (5) to zero as a necessary condition of solubility we get one algebraic equation of order n for γ.

For sake of simplicity we assume that we have n different solutions i.e. n eigenvalues with its n corresponding eigenvectors $\underline{x}_i$. Thus we do not consider solutions whose components are of the type $t^m \cdot \exp(-\gamma \cdot t)$, as discussed in every textbook on ordinary differential equations. If eigenvalues coincide,an appropriate minimal change of the system's parameters will make them different without changing the behaviour of the system within the time interval considered.

When we shall start below the other way round,as described in the introduction - i.e. by fitting experimental concentration data of an unknown system - we may always use two eigenvalues differing a little bit instead of coinciding eigenvalues, without changing the quality of the fit.

---

[1] These special solutions (3) do not necessarily correspond to possible real pharmacokinetic processes. In fact most of them do not, since they contain negative components in one or more compartments. Only one eigenvector is necessarily a real process: it's the eigenvector belonging to the smallest eigenvalue . After an injection or after the end of an infusion the full description of the system is a superposition of different eigenvectors with different eigenvalues or rate constants of disappearance and finally the term with the slowest rate constant becomes the leading term. If we wait long enough, only this leading term practically survives and we get it as a separate real process.

A full solution of (2) is now established by a superposition of the eigenvectors of type (3)

$$\underline{m}(t) = \sum_{i=1}^{n} c_i \underline{m}_i(t) = \sum_{i=1}^{n} c_i \underline{x}_i \exp(-\gamma_i t) \qquad (7)$$

and the coefficients c_i are to be adapted to fulfil the initial conditions describing the system's state at t = O, for instance the initial condition of intravenous injections of dose D into compartment 1, the central compartment.

$$\underline{m}(O) = \begin{pmatrix} D \\ 0 \\ 0 \\ \vdots \end{pmatrix} \qquad (8)$$

Inserting the general initial condition $\underline{m}$ (O) into equ. (7) we get

$$\underline{m}(O) = \sum_{i=1}^{n} c_i \underline{x}_i \qquad (9)$$

a system of linear equations for the c_i /cf. 8/.

<u>Eigenvector decomposition as an interpretative scheme</u>

In the preceding section we have used eigenvectors to solve a system of differential equations of a known compartment system. We now discuss in detail the two main aspects of pharmacokinetic eigenvector decomposition

i) the initial condition of an i.v.-injection into the central compartment as a distribution of the total dose D into the different eigenvectors.

ii) the special features of eigenvector dynamical description of a system.

i) Initial conditions.

The central injection is now interpreted as a way of excitation of the eigenvectors. By fixing an initial condition for the central component of each eigenvector, all other components in the lateral compartments are initiated as well, because of the fixed ratios of all components of an eigenvector. Of course the components of the different eigenvectors compensate to zero in the lateral compartments at t = O. These n-1 equations of compensation and the n-th condition that the total content of the central compartment amounts to D at t = O, fix the distribution of D within the central compartment.

Denoting the central content of the eigenvector i at t = O as d_i, we have

$$D = \sum_{i=1}^{n} d_i \qquad (1o)$$

This distribution is illustrated in fig. 1

Of course the total content of the eigenvector i, i.e. the sum

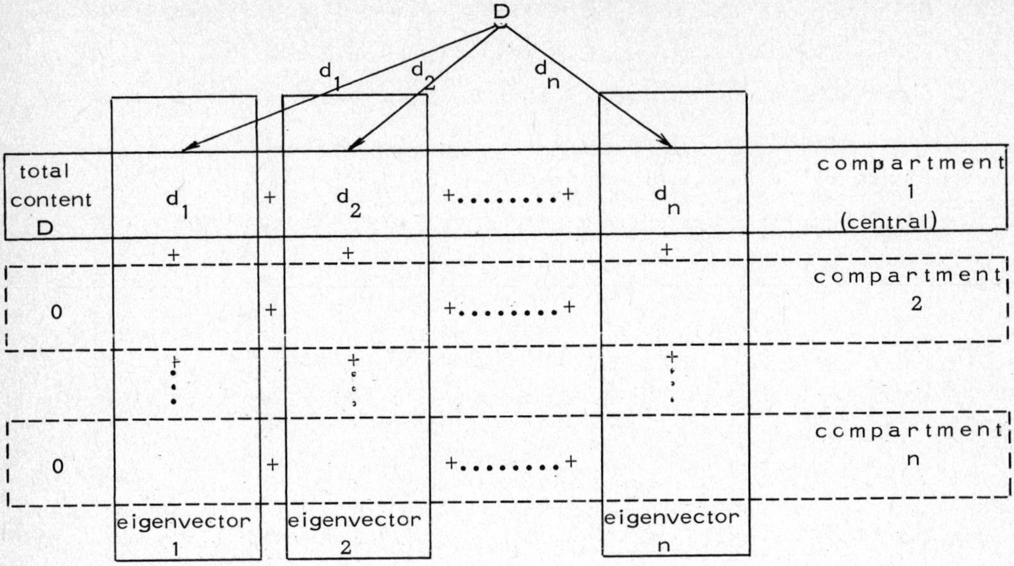

fig. 1
The intravenous injection as a distribution of the dose D among the central components d_i of the n eigenvectors at t=0. $D = \sum d_i$.

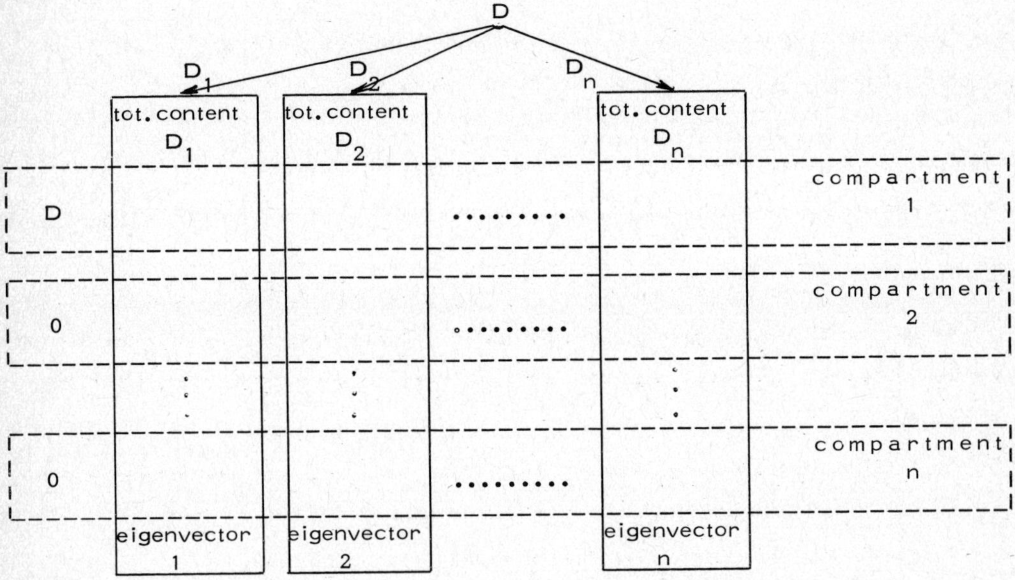

fig. 2
The intravenous injection as a distribution of the dose D among the total contents D_i of the n eigenvectors at t=0. $D = \sum D_i$.

of its components is not d_i. This total content D_i may be larger or smaller than d_i. For the eigenvector of the lowest eigenvalue - the so-called pseudo-distribution equilibrium /11/ - the total content is always larger than the central component, since all components are positive (cf. above). For other eigenvectors the sum D_i may be smaller than the central component d_i because of negative components and there must be some negative components since the total content of all eigenvectors amounts to D

$$\sum_{i=1}^{n} D_i = D \tag{11}$$

Since the pseudo-distribution equilibrium has $d_i < D_i$ there must be at least one eigenvector i with $d_i > D_i$.

We thus have two distributions of the same dose D

1) the distribution among the components d_i within the central compartment (fig. 1) and

2) the distribution among the total contents D_i of the eigenvectors (fig. 2).

ii) Eigenvector dynamics.

Eigenvector dynamics confines to the simple exponential decrease of each eigenvector determined by its eigenvalue. All exchange of drug between compartments is indirectly included in the exponential decrease. An exchange of substance between different eigenvectors does not occur by definition of the eigenvector: Each eigenvector is a separate mathematical solution of the system's dynamical equations.

Thus all transport phenomena between compartments are reduced to the fact that the ratios of different eigenvector components within the same compartment change because of their different eigenvalues, i.e. their different rate constants of decrease.

If we look at each eigenvector component, it behaves like a single compartment system with mono-exponential decrease. The peculiar feature of the eigenvector as a whole is the rigidity of the component's ratios.

A good illustration of eigenvector dynamics in a two-compartment-system has been presented by LEWI /12/.

Applications of eigenvector decomposition

Application 1 : Repeated dosage and accumulation

Every medical student learns how haert drugs like Digoxin or Digitoxin accumulate during repeated dosage. They learn the model of 1 compartment with one constant of elimination or one time constant of decreasing effect.

If the patient gets one pill per day, the drug will acculul ate up to a saturation level determined by the balance

$$\begin{pmatrix}\text{saturation}\\ \text{level}\end{pmatrix} \cdot \begin{pmatrix}\text{loss of effect}\\ \text{per day}\end{pmatrix} = \begin{pmatrix}\text{absorbed dose}\\ \text{per day}\end{pmatrix} \qquad (12)$$

The blood curve is periodic if the dose per day just compensates for the decrease per day.

If we take into account more than one compartment, i.e. more than one time constant, we may either ask for the drug content in the central compartment or in the whole body. For the two-compartment-system this is already a full description of the system, since the content of the single lateral compartment is merely the difference between the total and the central content.

In order to get the accumulation in the central compartment, we need a curve of the central concentration or amount of one single dosage decomposed into its components with their different time constants

$$m_1 (t) = \sum_{i=1}^{n} d_i \exp (- \gamma_i t) \qquad (13)$$

Because of the superposition principle we may calculate the accumulation of each central component separately in the same way as we do in a single compartment system. Having done the accumulation calculation for the n central components separately we finally sum up the results to get the total result for the central compartment.

The central supposition of this strategy is the superposition principle to justify the conclusion from the single dose effect to repeated dose effect without considering the proceedings in the lateral compartments. A supposition of second importance is that of exponential decomposition which simplifies the accumulation calculations.

In order to get the total dose retained in the system during repeated dosage, we need an additional information, i.e. the relation between the central component of an eigenvector and its total drug content. We get this additional information by the supposition of central elimination (cf. introduction) as follows:

Provided that we know the rate constant of final elimination k_{10} (we shall get it below from our central data), we may calculate the fraction of dose excreted during the time interval dt from the eigenvector i (central component: $d_i \cdot \exp(-\gamma_i t)$) by

$$k_{1o} \cdot d_i \cdot \exp(-\gamma_i t)\, dt \tag{14}$$

Integrating (14) from zero to infinity, we get the total content D_i of the i-th eigenvector

$$D_i = \int_0^\infty k_{1o}\, d_i \cdot \exp(-\gamma_i t)\, dt \tag{15}$$

$$D_i = d_i \cdot \frac{k_{1o}}{\gamma_i} \qquad i = 1 \ldots n \tag{16}$$

In order to get k_{1o} we may sum up the fractional doses (16) to the total dose according to (11).

$$D = \sum_{i=1}^{n} d_i \cdot \frac{k_{1o}}{\gamma_i} \tag{17}$$

or

$$k_{1o} = \frac{D}{\sum_{i=1}^{n} \frac{d_i}{\gamma_i}} \tag{18}$$

which is a special way of writing 'DOST's principle' /1, 24/

$$k_{1o} = \frac{D}{\int_0^\infty m_1(t)\, dt} \tag{19}$$

Since we thus have by (16) the conversion between d_i and D_i we may calculate accumulation on the basis of the single dose time course

$$D_i \exp(-\gamma_i t) \quad \text{or} \quad d_i \cdot \frac{k_{1o}}{\gamma_i} \cdot \exp(-\gamma_i t) \tag{20}$$

for the i-th eigenvector in the same way as we stated for the central component or a single compartment system and again we get the final results by summing up all eigenvectors. Some detailed formula are given in /3/.

<u>Application 2: the volume of distribution</u>

For a single compartment system we simply get the volume of distribution by applying a dose D to the single compartment and measuring the concentration y (O) extrapolated back to application time t = O

$$V = \frac{D}{y(0)} \tag{21}$$

For more than one compartment we have to make sure that the lateral compartments have the same concentration as the central one i.e.

i) we have to apply an infusion until equilibrium is reached,

ii) the infusion must be applied to the compartment of elimination.

Thus we do not consider systems with elimination from different compartments, which need either somehow arbitrary extensions of equilibrium formula or separate determination of compartment volumes / cf. 4 p.200-211/.

We now start our calculations from the concentration versus time curve of a single i.v.-injection

$$y_1(t) = \frac{m_1(t)}{V_1} = \sum_{i=1}^{n} \frac{d_i}{V_1} \exp(-\gamma_i t) \tag{22}$$

with V_1 being the volume of the central compartment /13/

$$V_1 = \frac{D}{y_1(0)} \tag{23}$$

We assume central elimination, i.e. (16).

It is possible to assign a volume to an eigenvector by extending the general formula for the volume of distribution / cf. 4 p. 2o9/

$$V = \frac{m^*}{y^*} = V_1 \cdot \frac{m^*}{m_1^*} \tag{24}$$

$$\begin{pmatrix}\text{volume of}\\ \text{distribution}\end{pmatrix} = \begin{pmatrix}\text{central}\\ \text{volume}\end{pmatrix} \cdot \left(\frac{\text{tot. drug content of the system at infusion equilibrium}}{\text{central content at infusion equ.}}\right) \tag{24a}$$

The star generally denotes infusion equilibrium.

Our definition of the volume V' of an eigenvector is

$$V_i' = V_1 \cdot \frac{D_i^*}{m_1^*} \qquad i = 1, \ldots, n \tag{25}$$

where D_i^* is the total drug content of an eigenvector at infusion equilibrium.

By summing up the volumes of the different eigenvectors we shall eventually get the total volume of distribution aggreeing with (24)

$$V = \sum_{i=1}^{n} V_i' \tag{26}$$

In order to evaluate (25) we first replace m_1^* via

$$m_1^* \cdot k_{1o} = \dot{D} \tag{27}$$

where $\dot{D}$ is the infused dose per time belonging to the central equilibrium drug content m_1^* and (27) states the equilibrium balance from which we get

$$m_1^* = \dot{D} / k_{1o} \tag{28}$$

If we denote by $\dot{D}_i$ the dose per time infused into the i-th eigenvector and if we set γ_i in analogy to k_{1o} and D_i^* an analogy to m_1^*,
we get the equilibrium balance

$$D_i^* \cdot \gamma_i = \dot{D}_i \tag{29}$$

or

$$D_i^* = \dot{D}_i / \gamma_i \tag{3o}$$

similarly to (27) and (28) adapted to the i-th eigenvector as a single compartment system with the elimination constant γ_i. Inserting (28) and (3o) into (25) we get

$$V_i' = V_1 \frac{\dot{D}_i}{\dot{D}} \cdot \frac{k_{1o}}{\gamma_i} \tag{31}$$

It remains to eliminate $\dot{D}_i / \dot{D}$, i.e. the fraction of infusion entering the total i-th eigenvector. Since an infusion is mathematically nothing but a series of infinitesimal injections the infusion fraction equals the injection fraction D_i/D entering the total i-th eigenvector, which may be expressed via (16) in terms of injection central data:

$$\dot{D}_i / \dot{D} = D_i/D = \frac{d_i}{D} \cdot \frac{k_{1o}}{\gamma_i} \tag{32}$$

Inserting (32) into (31) we have

$$\boxed{V_i' = V_1 \cdot \frac{d_i}{D} \cdot \left(\frac{k_{1o}}{\gamma_i}\right)^2} \qquad i = 1, .., n \tag{33}$$

the final formula for the volume of the i-th eigenvector. The summation (26) eventually provides

$$V = V_1 \cdot \sum_{i=1}^{n} \frac{d_i}{D} \cdot \left(\frac{k_{1o}}{\gamma_i} \right)^2 \tag{34}$$

<u>A footnote on concentrations versus amounts and the related volumes</u>

Sometimes the variables of the compartmental equations are interpreted not as amounts as in (1') but as concentrations / cf. 1/. In this case all amounts had been devided by the volume of the central compartment, not by the proper volume of distribution of its compartment. By this means the simple structure of the coefficients as stated for k_{jj} following formula (1) is retained.

The drawback of this procedure is that false interpretations of concentration gradients are likely to occur. Pharmacokinetic compartmental systems for instance do not allow a netto flux of drug against a concentration gradient if proper concentrations are considered.

Application 3 : Absorption analysis

I do not want to get into details of wellknown strategies of absorption analysis /15, 16, 17, 18/ . Some recent applications are /19, 20, 21/. The first-pass-effect /22, 23, and indirectly 24/ is not considered here.

The strategy proposed in this section, is a type of deconvolution /10, 25/ but it does not use Laplace-Transformation. Like the analysis of SCHOLER and CODE /15/ it does not leave the field of concentration versus time curves.

Our basic approach is the following. We assume to have the two informations quoted below:

i) the central amount (or concentration) versus time curve after intravenous injection and its decomposition into exponentials

ii) the central amount (or concentration) curve of the absorption process $m_1(t)$

(Central concentrations and amounts may be converted into each other via the central volume (23). We confine to amounts)

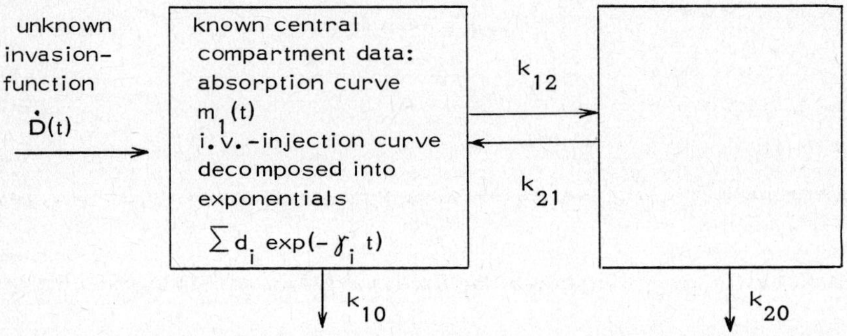

fig. 3
Sample of an absorption process. The k's are assumed to be unknown.

Our aim is to get the invasion function $\dot{D}(t)$.

The change of the central drug amount dm_1/dt is a superposition of 3 effects

(i) the unknown input $\dot{D}(t)$

(ii) the transport into and back from the lateral compartments

(iii) the elimination from the central compartment.

In order to get (i) separated we shall decompose $m_1(t)$ into its central components belonging to the n different eigenvectors

$$m_1(t) = \sum_{i=1}^{n} m_{1i}(t) \qquad (35)$$

Summing up all exponential decrease according to the processes (ii) and (iii) we get the total netto efflux from the central compartment as

$$\sum_{i'=1}^{n} \gamma_{i'} \cdot m_{1i'}(t) \tag{36}$$

Thus the overall balance for the central compartment is given by equating the input $\mathring{D}(t)$ to the change of the central content plus the netto central efflux

$$D(t) = \frac{dm_1(t)}{dt} + \sum_{i'=1}^{n} \gamma_{i'} \, m_{1i'}(t) \tag{37}$$

This balance does not yet determine the components $m_{1i}(t)$. We still need the prescription, how the total input $\dot{D}(t)$ is distributed among the different eigenvectors. This distribution rule is the same as with i.v.-injection (cf. fig. 1 and formula (10)), i.e. it follows from the decomposition of the i.v.-injection curve. Thus the input into the central component of the i-th eigenvector is given by

$$\frac{d_i}{D} \mathring{D}(t) = \frac{d_i}{D} \left\{ \frac{dm_1(t)}{dt} + \sum_{i'=1}^{n} \gamma_{i'} \, m_{1i'}(t) \right\} \tag{38}$$

according to (37) and the total balance for $m_{1i}(t)$ is

$$\frac{dm_{1i}(t)}{dt} = \frac{d_i}{D} \left\{ \frac{dm_1(t)}{dt} + \sum_{i'=1}^{n} \gamma_{i'} \, m_{1i'}(t) \right\} - \gamma_i \, m_{1i}(t) \qquad i=1\ldots n \tag{39}$$

which is a system of n linear differential equations for the n unknown functions $m_{1i}(t)$. The inhomogeneity or the driving force is given by $\frac{d_i}{D} m_1(t)$ and the initial values are zero for all $m_{1i}(t)$ at t=0, i.e. $m_{1i}(0)=0$ at the begin of the absorption process.

To calculate the $m_{1i}(t)$ we start by solving the homogeneous system belonging to (39) i.e.

$$\frac{d\, m^{\delta}_{1i}(t)}{dt} = \frac{d_i}{D} \sum_{i'=1}^{n} \gamma_{i'} \, m^{\delta}_{1i'}(t) - \gamma_i \, m^{\delta}_{1i}(t) \qquad i=1\ldots n \tag{40}$$

$$m^{\delta}_{1i}(0) = \frac{d_i}{D} \qquad i=1\ldots n \tag{41}$$

The index δ denotes that these initial conditions describe a δ-type inhomogeneity, i.e. an instantneous injection, and the final solution for $m_{1i}(t)$ will be a superposition of all responses of the inhomogeneity $m_1(t)$ being decomposed into infinitesimal injections at different times τ

$$m_{1i}(t) = \int_0^t m_1(\tau) \, m^{\delta}_{1i}(t-\tau) \, d\tau \qquad i=1\ldots n \tag{42}$$

i.e. a convolution of the dimensionless $m^{\delta}_{1i}(t)$ and $m_1(t)$.

We'solve' (40)/(41) by a simple interpretation; we go back from the system of differential equations to a corresponding compartmental system.

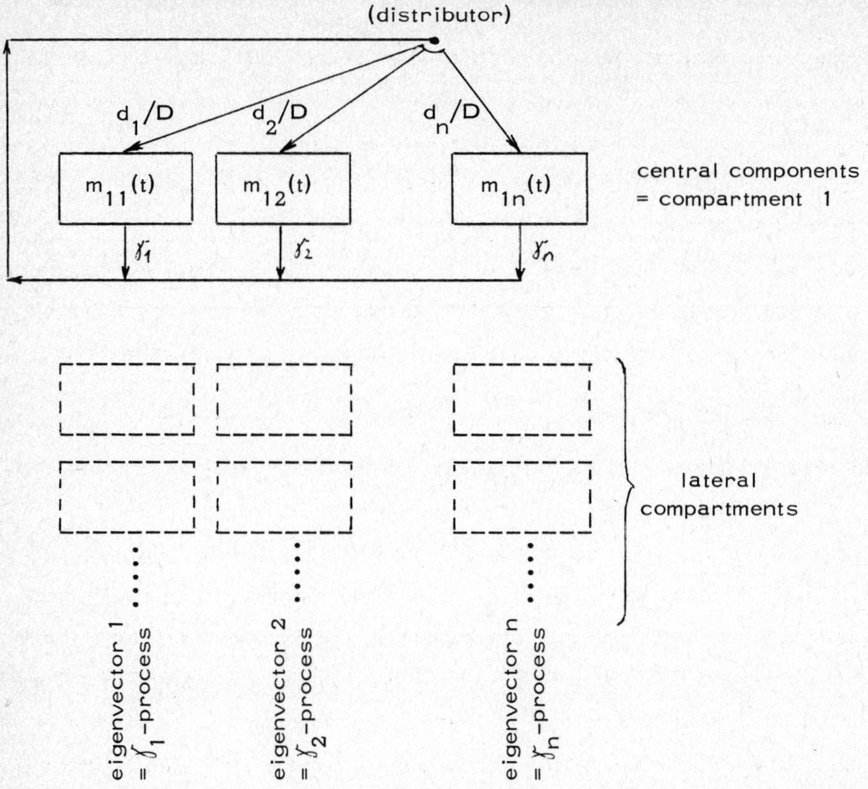

fig. 4

We associate the n central components $m_{1i}(t)$ with n separate compartments, each having its elimination constant γ_i, and the efflux of all these compartments is collected and fedback into the n compartments according to the distribution d_i/D (fig. 4). This circular process does not loose any substance, which immediately results in a crucial statement on the new eigenvectors (in order to distinguish between the eigenvalues γ_i of our original system and the eigenvalues of (40), we call the latter ones 'new eigenvalues'):

either the new eigenvalue is zero

or the total content of the new eigenvector must be zero

$$\sum_{i=1}^{n} m_{1i}(t) = 0 \tag{43}$$

since a non-zero eigenvalue will decrease each component, i.e. change the content of the system.

The new eigenvector for the new eigenvalue zero is immediately found as

$$m_{1i}(t) = \frac{d_i}{D} \cdot \frac{1}{\gamma_i} \qquad i=1\ldots n \tag{44}$$

the time-independent stationary solution of (40).

Discussing the central feedback-model of fig. 4, we forgot up to now about the lateral compartments of our original model, implied in fig. 4 as dashed lines; but we know that there are fixed ratios of the central component of each eigenvector and the other components in the lateral compartments. We shall use this fact now to interprete the central process in terms of the whole original system.

Concerning the non-zero new eigenvalues, we interprete the difference between (40) and the original system thus that the latter has been supplemented by an infusion which keeps the central content at zero level. Since the supplementing term in (40)

$$\frac{d_i}{D}\sum_{i'=1}^{n} \gamma_{i'} \cdot m_{1i'}(t) \qquad (45)$$

has a distribution of type d_i/D, this interpretation as an infusion is legal. The rest of (40) is the familiar eigenvector decreasing term.

In consequence of replacing thus the absorption invasion function $\dot{D}(t)$ by a compensating infusion, the system behaves as if the central compartment with its constantly zero content is cut off. The remaining system of n-1 compartments has n-1 new non-zero eigenvalues belonging to (40). If we have a mamillary system (i.e. a system where all lateral compartments are directly attached to the central one without additional connexions between each other) these eigenvalues are the total efflux constants of the n-1 lateral compartments, i.e. $k_{i1}+k_{i0}$. Since eigenvector decomposition in practice will not make any sense for n greater than 3, this truncated system is always very simple.

The new eigenvectors belonging to the new eigenvalues are interpreted now in a simililarly intuitive way as we got the new eigenvalues. We only decompose the eigenvectors of the truncated system in terms of the original eigenvectors to get the corresponding central components. For mamillary systems these eigenvectors of the truncated system are injections of drug into one single lateral compartment. Just like the central injection led to initial conditions where all compartments but the central one had zero content, now all compartments but a lateral one have zero content, and this initial condition has to be decomposed in terms of the original eigenvectors in order to get the central components of the new eigenvectors, i.e. the real new eigenvectors which are distributions of central components.

The decomposition of the initial condition (41) in terms of the new eigenvectors gives us the time course of all $m^{\delta}_{1i}(t)$ as a sum of exponentials, the new eigenvectors being the time constants of decrease. We than proceed by the convolution (42) to the n $m_{1i}(t)$ and by (35) to the invasion function $D(t)$, our final result.

All integrals may be calculated readily, since they contain nothing but exponentials.

The final result will eventually contain nothing but the experimental data i) and ii) from the central compartment, quoted at the beginning of this section. In order to interprete the intermediate steps of our calculation, we referred to the original compartment system which was supposed to be unknown. But it is not necessary to analyse the system, in order to solve (40)/(41), and in general it is impossible to get a full analysis of the system from the information i) and ii) above. Our straightforward mathematical solution of (40)/(41) only includes indirectly some system analysis; in case we want to refer to special types of compartment models, we immediately have an interpretation of our model independent results in terms of the special models, we now refer to.

It should be kept in mind that all calculations above admit most simple intuitive interpretations if we refer to mamillary systems.

Final remarks

The eigenvector decomposition principle is a special way to look at transport phenomena in multi-compartment systems. Instead of asking for the time course of drug amount in each compartment, we do all calculations somehow perpendicular to this point of view: from all compartments we comprehend the entities of all components belonging to the same time constant γ_i, i.e. to the same eigenvalue of the system. These entities are the eigenvectors or γ-processes (to signify the common time constant of such a process).

Being acquainted with the eigenvector decomposition is mainly a feeling for two facts:

i. for the rigidity of the constant ratios of all components belonging to the same time constant γ_i

ii. for the fact that there is no exchange of substance between different eigenvectors.

Applying the eigenvector decomposition is mainly a full use of linearity in the following manner: you proceed from the compartmental coordinate system to the eigenvector system; you do all your calculations in this system without worrying about negative components, despite of psychological inhibition; you may finally return to the compartmental coordinate system if necessary.

Eigenvector decomposition especially suites model independent analysis of central compartment data. Of course it cannot extract exact information from noisy data. In practice it does not make any sense to extract more than 2 or 3 exponentials from experimental data.

Eigenvector decomposition sometimes admits model independent derivations. Of course a more transparent proof will better point out the weakness of its suppositions, instead of sweeping it under the carpet of complicated special calculations. It may form the basis of a more complicated approach. This aspect is as well a chance and a temptation. Beware of PETER's principle of biomathematics. Let me explain: This principle of PETER and HULL originally describes the trend to climb up a carreer until the climber gets incompetent and unhappy. In biomathematics there is a corresponding temptation to climb up the tree of complexity until the weakness of the approach is no more accessible.

References

/1/ Dost, F.H.: Grundlagen der Pharmakokinetik. 2nd edition, Thieme Stuttgart 1968.

/2/ Thron, C.D.: Linearity and superposition in pharmacokinetics. Pharmacol. Rev. 26, 3-31(1974).

/3/ Müller-Schauenburg, W.: A new method for multi-compartment pharmacokinetic analysis: the eigenvector decomposition principle. Europ.J.Clin.Pharmacol. 6, 203-206(1973).

/4/ Riggs, D.S.: The mathematical approach to physiological problems. M.I.T.Press paperback, Cambridge Mass., London 1970.

/5/ Rockoff, M.L.: Interpretation of the clearance curve of a diffusible tracer by blood flow in terms of a parallel-compartment model. p.108-126 in Computer processing of dynamic images from an Anger scintillation camera. K.B.Larson and J.R.Cox ed. The society of nuclear medicine, Inc.New York 1974

/6/ Simon, W.: A method of exponential separation applicable to small computers. Phys.Med.Biol. 15, 355-360(1970)

/7/ Glass, H.I. and A.C.DeGarreta: The quantitative limitations of exponential curve fitting. Phys.Med.Biol. 16, 119-130(1971)

/8/ Feldmann, U.: Lecture notes in this volume.

/9/ Schneider, B.: Mathematical and statistical problems in pharmacokinetics. Advances in the Bioschiences. Schering Workshop on pharmacokintics, Berlin, May 8and 9, 1969. Pergamon Press - Vieweg 1970, p.103-112.

/10/Rescigno, A. and G. Segre: Drug and tracer kinetics. Blaisdell, Waltham, Mass. 1966.

/11/Gibaldi, M., R.Nagashima and G.Lewy: Relations between drug concentration in plasma or serum and amount of drug in the body. J.Pharm.Sci. 58, 193-197(1969).

/12/Lewi, P.J.: Pharmacokinetics and eigenvector decomposition. Two-compartment open system after rapid intravenous administration. Arch.int.de Pharmacodynamie et de Therapie 215, 283-300(1975).

/13/ Gladtke, E. und H.M.v. Hattingberg: Pharmakokinetik. Springer, Berlin Heidelberg New York 1973.

/14/ Dost, F.H.: Absorption, Transit, Occupancy und Availments als neue Begriffe in der Biopharmazeutik. Klin. Wschr. 50, 410-412(1972).

/15/ Scholer, J.F. and C.F. Code: Rate of absorption of water from stomach and small bowel in human beings. Gastroenterology 27, 565-577(1954).

/16/ Wagner, J.G. and E. Nelson: Percent absorbed time plots derived from blood level and/or urinary excretion data. J. Pharm. Sci. 52, 610-611(1963).

/17/ Loo, J.C.K. and S. Riegelman: New method for calculating the intrinsic absorption rate of drugs. J. Pharm. Sci. 57, 918-928(1968).

/18/ Kwan, K.C. and A.E. Till: Novel method for bioavailability assessment. J. Pharm. Sci. 62, 1494-1497(1973).

/19/ Wagner, J.G.: Application of the Wagner-Nelson absorption method to two-compartment open model. J. Pharmacokinetics Biopharmaceutics 2, 469-486 (1974).

/20/ Wagner, J.G.: Application of the Loo-Riegelman absorption method. J. Pharmacokinetics Biopharmaceutics 3, 51-67(1975)

/21/ Till, A.E., L.Z. Benet and K.C. Kwan: An integrated approach to the pharmacokinetic analysis of drug absorption. J. Pharmacokinetics Biopharmaceutics 2, 525-544(1974)

/22/ Gibaldi, M., R.N. Boyes and S. Feldman: Influence of first-pass effect on availability of drugs on oral administration. J. Pharm. Sci. 60, 1338-1340(1971).

/23/ Rowland, M.: Influence of route of administraion on drug availability. J. Pharm. Sci. 61, 70-74(1972).

/24/ Nüesch, E.: Proof of the general validity of Dost's law of corresponding areas. Europ. J. clin. Pharmacol. 6, 33-43(1973).

/25/ Silverman, M. and A.S.V. Burgen: Application of analogue computer to measurement of intestinal absorption rates with tracers. J. Appl. Physiol. 16, 911-913(1961).

A General Approach to Multicompartment Analysis and Models for the Pharmacodynamics

U. Feldmann and B. Schneider
Department of Biometrics and Medical Informatics
Medical School Hannover

Contents

Introduction

In this paper concepts for mathematic models in pharmacokinetics and pharmacodynamics are discussed. As far as kinetic reactions of the first or zero order are considered, a complete linear theory can be introduced for pharmacokinetics.
The mathematic treatment of pharmacodynamic leads to nonlinear dose-effect relationships and time-effect relationships, the general explicit representations of which are difficult, often not possible.

The concepts "pharmacokinetics" and "pharmacodynamics" are defined in accordance with the PHARMA-KODEX.
Pharmacokinetics is understood to refer to the behaviour of a drug in the organism, namely resorption, distribution, biochemical transformation (metabolism) and excretion.
Pharmacodynamics is understood to refer to the medicinally caused alterations of the normal or experimentally altered functions of the organism. The results must thereby be described, first in quantitative form (dose-effect curve, time-effect curve, etc.) and then - whenever possible - in comparison with materials whose effects are well known.

Our goal is to describe adequately the distribution and effect of a substance in the human organism.
Finally the descriptive system in which these processes can be represented is mathematics.
A mathematisation presupposes a scientific knowledge of the chemico-physical characteristics of the observed biological systems. Due to the complexity of biological processes, such knowledge will not be complete but rather will precipitate a biological model, which seeks to clarify the essential characteristics of the biological system and which is unambiguously describable with mathematic structures.
The question whether a model is optimal can not be determined within mathematics, as it originates in a different frame of reference.
We recognize reality through measurements on the organism. The adequacy of a mathematical model, which describes this reality, is not solely determined by how well the measurements on the system are fitted by the calculations in the model.
It is more important that the parameters of the mathematical model be biologically interpretable and experimentally reproducible. It should therefore be the goal of a model approach to explain as far as is possible the biological variation of reality using only a few parameters, i.e. to reduce the copious experimental data to a few independent statements.

Mathematical models are used for the simulation and approximation of biological systems.
In the simulation method behaviour of the system is calculated using predetermined parameters.
In the approximation method the parameters of the model are estimated from the experimental data. When the estimates are based on mathematical-statistical methods, it is possible to determine the degree of variation in the parameters and thus to make statements about the relevance of the parameters within a model.

Part I: A General Approach to Multicompartment Analysis

1. Multicompartment Models in Pharmacokinetics

We understand compartments to refer to the real or ficticious distribution spaces of the medication in the organism.

Thus the blood circulatory system, the liver, or the intestinal tract are viewed as real compartments and the plasma or serum as ficticious ones.

We assume that medications disperse directly and spatially homogenously in the compartments.

In a compartment model, a temporal change in the quantity of medication should only be possible as a result of interdependent reaction between the compartments.

It is then assumed that the change in drug quantity is proportional to the drug quantity present in a compartment; thus only kinetic reactions of the first order are observed:

$$\dot{m}(t) = -k \cdot m(t) \tag{1.1}$$

$m(t)$ [mass] is the drug quantity at time t;

k [1/time] is the relative reaction speed, which is taken as a temporal and spatial constant; and

$\dot{m}(t)$ [mass/time] is the temporal change in the drug quantity.

The following three-compartment model, consisting of an application compartment, a dispersal compartment, and an effect compartment, can be viewed as a basic model for pharmacokinetics.

Model 1.1: Basic Model of Pharmacokinetics
Exemplified by Coronary Glycosides

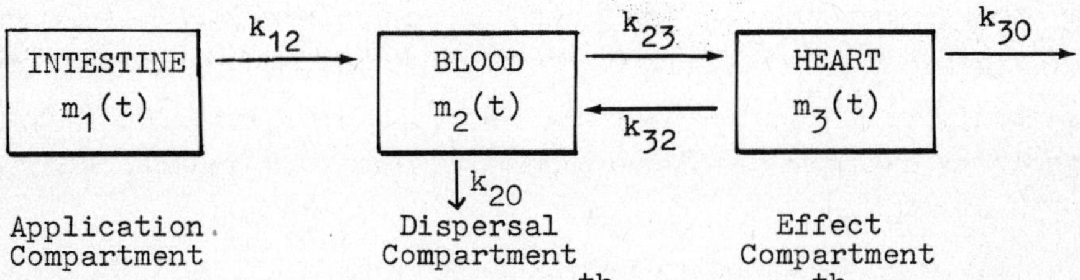

k_{ij} are the transition rates from the ith into the jth compartment.
k_{io} represents the transition from the ith compartment into the system exterior.

The associated differential equations follow from the principle expressed in (1.1):

$$\begin{aligned} \dot{m}_1(t) &= -k_{12}\, m_1(t) & m_1(0) &= d_1 \\ \dot{m}_2(t) &= -(k_{23}+k_{20})m_2(t)+k_{12}m_1(t)+k_{32}m_3(t) & m_2(0) &= 0 \\ \dot{m}_3(t) &= -(k_{32}+k_{30})m_3(t)+k_{23}m_2(t) & m_3(0) &= 0 \end{aligned} \tag{1.2}$$

While (1.1) can still be accepted as a special case of Fick's Law, (1.2) no longer obeys the law of diffusion but rather also describes

the active transportation of medications.
Model (1.2) proceeds from a single oral application with dose d_1 under the assumption that there are no glycosides in the blood circulatory system or in the heart at time t=0.
A general n-compartment system which describes kinetic reactions of the first order can thus be represented by the differential equation system:

$$\dot{m}_i(t) = -\sum_{\substack{j=o \\ j\neq i}}^{n} k_{ij}\, m_i(t) + \sum_{\substack{j=1 \\ j\neq i}}^{n} k_{ji}\, m_j(t) \tag{1.3}$$

with the initial conditions $m_i(0) = d_i$ for a single application of dose d_i in the ith compartment with i=1,2,...,n.
In the literature (Sheppard-Householder 1951, Berman-Schoenfeld 1956, Rescigno-Segre 1966) the Laplace transformation is used in the analysis of such multicompartment systems. The advantage of this transformation lies primarily in the fact that differentiation and integration in the Laplace space are reduced to multiplication and division. This transformation has proved itself primarily in the treatment of boundary value problems.
It is, however, possible with the aid of spectral decomposition and linear algebra to prefer a direct explicit solution for the representation of multicompartment models in pharmacokinetics, which are described mathematically by homogenous or inhomogenous linear initial-value differential equation systems of the first order.
The general method presented here for solving linear differential equations with initial-value problems is mathematically familiar ; it appears, however, not to have been used as yet in pharmacokinetics. It presents a direct, lucid generalisation of a one-compartment model.

Primarily it allows the various forms of application of medications common in medicine to be described uniformly in general multicompartment models.

2. Single Application of Medications

A single application of medication can occur in oral form and, as an injection, intravenously, subcutaneously and intramuscularly. We assume that at time t=0 the dose d_i has been applied by a single application of medication in the ith compartment (i=1,2,...,n).

The general n-compartment system (1.3) is then representable in matrix form as

(2.1)* $$\begin{pmatrix} \dot m_1(t) \\ \dot m_2(t) \\ \vdots \\ \dot m_n(t) \end{pmatrix} = \begin{pmatrix} -k_{11} & k_{21} \cdots k_{n1} \\ k_{12} & -k_{22} \cdots k_{n2} \\ \vdots & \vdots \quad \vdots \\ k_{1n} & k_{2n} \quad -k_{nn} \end{pmatrix} \cdot \begin{pmatrix} m_1(t) \\ m_2(t) \\ \vdots \\ m_n(t) \end{pmatrix} \text{ and } \begin{pmatrix} m_1(0) \\ m_2(0) \\ \vdots \\ m_n(0) \end{pmatrix} = \begin{pmatrix} d_1 \\ d_2 \\ \vdots \\ d_n \end{pmatrix}$$

in which

(2.2) $$k_{ii} = \sum_{\substack{j=0 \\ i \neq j}}^{n} k_{ij} \quad \text{and} \quad 0 \leq k_{ij} < 1 \quad \text{for} \quad \begin{array}{l} i=1,2,\ldots,n \\ j=0,1,2,\ldots,n \end{array}$$

(2.2)* Furthermore $0 < k_{ii} < 1$

Formally (2.1)*can be summarized as

(2.1) $$\boxed{\dot m(t) = K \cdot m(t) \quad \text{and} \quad m(0) = d}$$

when $\dot m(t)$, $m(t)$ and d are vectors, i.e. $(n \times 1)$ matrices and K is an $(n \times n)$ matrix.

A partial solution of (2.1) is determinable by the mathematically familiar method of characteristic equation with exponentials.

(2.3) $$m(t) = X_i \cdot e^{\lambda_i t}$$

when $X_i = (x_{1i}, x_{2i}, \ldots, x_{ni})'$ is an $(n \times 1)$ matrix and λ_i is a real or complex number.

If the expression (2.3) is inserted into (2.1) then the formula

(2.4) $$K \cdot X_i = X_i \cdot \lambda_i \qquad (i=1,2,\ldots,n)$$

appears as a necessary condition for X_i and λ_i.

The determination of X_i and λ_i thus leads to an eigenvalue problem, in which λ_i is the eigenvalue and X_i is the associated (right-hand) eigenvector. Since the matrix K is real but in general not symmetrical, complex eigenvalues can also occur.

Provided (2.2), it can be shown (Berman-Schoenfeld 1956) that the real part of the eigenvalues are always less or equal zero and that the imaginary parts only occur when the real part is less than zero; thus

(2.5) $$\begin{array}{ll} \mathrm{Re}(\lambda_i) \leq 0 & \text{and} \\ \mathrm{Im}(\lambda_i) \neq 0 & \text{only if } \mathrm{Re}(\lambda_i) < 0 \end{array}$$

We consider the eigenvectors X_i as column vectors of an $(n \times n)$ matrix X and the eigenvalues λ_i as a diagonal matrix Λ, then (2.4) can be represented by

(2.6) $$K \cdot X = X \cdot \Lambda$$

in which

$$X = \begin{pmatrix} x_{11} & x_{12} & \cdots & x_{1n} \\ x_{21} & x_{22} & \cdots & x_{2n} \\ \vdots & \vdots & & \vdots \\ x_{n1} & x_{12} & \cdots & x_{nn} \end{pmatrix} \quad \text{and} \quad \Lambda = \begin{pmatrix} \lambda_1 & 0 & \cdots & 0 \\ 0 & \lambda_2 & \cdots & 0 \\ \vdots & \vdots & & \vdots \\ 0 & 0 & & \lambda_n \end{pmatrix}$$

Assuming that all eigenvalues are unequal zero and different from each other, then the eigenvectors are linearly independent and the determinant of the matrix X differs from zero. This is the case which we wish to continue to consider.
Through (2.3) n partial solutions to the problem (2.1) have been found, which are sufficient for the differential equations but not for their initial conditions. The general solution can be derived from the partial solutions according to the familiar superposition principle of differentiation and results in

$$m(t) = X \cdot e^{\Lambda t} \cdot X^{-1} \cdot d \tag{2.7}$$

There we have, written in matrix form, a direct solution of the general n-compartment model with a single application of medication. $e^{\Lambda t}$ is an (n x n) diagonal matrix of the form

$$e^{\Lambda t} = \begin{pmatrix} e^{\lambda_1 t} & 0 & \cdots & 0 \\ 0 & e^{\lambda_2 t} & \cdots & 0 \\ \vdots & \vdots & & \vdots \\ 0 & 0 & \cdots & e^{\lambda_n t} \end{pmatrix}$$

It is consequently possible, for fixed, predetermined parameters k_{ij} and initial doses d_i, to simulate the temporal curve of the amount of medication in all compartments by simple matrix operations.
The thus explicit solution of multicompartment models leads to sums of exponentials, in which damped oscillations can occur in certain circumstances, but in which pure oscillations are impossible.
In medical practice one would like to estimate the parameters k_{ij} from experimental data. It is even often the case, for example in the clinical bromsulphalein test, that measurements can only be taken in one compartment, and the behaviour of another compartment will be deduced from these measurements.
It is clear that, in general, when taking measurements on a single compartment of an n-compartment system, no all n^2 parameters and n initial values can be estimated simultaneously.
If the kth compartment is the application compartment, such that $d_k > 0$ and $d_j = 0$ for $j \neq k$ and if the ith compartment is the measured compartment, then the exponential trial function

$$m_i(t) = \sum_{j=1}^{n} a_{ji} e^{-\gamma_j t} \tag{2.8}$$

is selected for the approximation, and a_{ji} and γ_j as well as their asymptotic standard deviations are numerically determined with the aid of the maximum likelihood method.

The question arises, whether d_k and k_{ij} are calculable from the estimated values a_{ji} and γ_j.

Now let $X = \{x_{ik}\}$, $X^{-1} = \{z_{ik}\}$ and $\lambda_j = -\gamma_j$

then, given the just-mentioned preconditions

$$(2.9) \qquad m_i(t) = d_k \cdot \sum_{j=1}^{n} x_{ij} \cdot z_{jk} \cdot e^{-\gamma_j \cdot t}$$

follows from (2.7).

The relationship between the sums of the exponentials (2.8) and the compartment model (2.1) is thus given by

$$(2.10) \qquad a_{ji} = d_k \cdot x_{ij} \cdot z_{jk} \quad \text{and} \quad \gamma_j = -\lambda_j$$

1. Example: Bromsulphalein Test

Given the two-compartment model

Model 2.1: Compartment Model of the Clinical Bromsulphalein Test

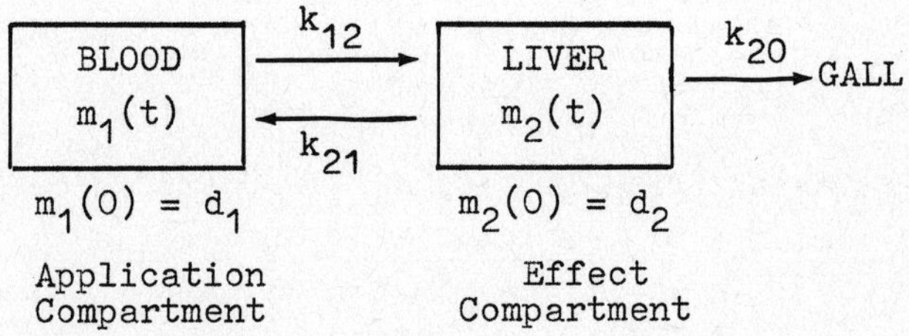

The differential equation for this model is

$$(2.11) \qquad \begin{aligned} \dot m_1(t) &= -k_{12}m_1(t) + k_{21}m_2(t) \quad &; m_1(0) = d_1 \\ \dot m_2(t) &= +k_{12}m_1(t) - (k_{21}+k_{20})m_2(t) \quad &; m_2(0) = d_2 \end{aligned}$$

in which

$$K = \begin{pmatrix} -k_{12} & k_{21} \\ k_{12} & -(k_{21}+k_{20}) \end{pmatrix}$$

The eigenvalues λ_i of K are

$$\lambda_{1/2} = -\tfrac{1}{2}(k_{12}+k_{21}+k_{20}) \pm \sqrt{\tfrac{1}{4}(k_{12}+k_{21}+k_{20})^2 - k_{12} \cdot k_{20}}$$

with $\lambda_1 \leq \lambda_2 < 0$

Furthermore the matrix of the (right-hand) eigenvectors and its inverse are

$$X = \begin{pmatrix} -(\lambda_1+k_{20}) & -(\lambda_2+k_{20}) \\ \lambda_1 & \lambda_2 \end{pmatrix}$$

$$X^{-1} = \frac{1}{k_{20}(\lambda_1 - \lambda_2)} \begin{pmatrix} \lambda_2 & (\lambda_2+k_{20}) \\ -\lambda_1 & -(\lambda_1+k_{20}) \end{pmatrix}$$

Now let $\lambda_1 \neq \lambda_2$; $\gamma_i = -\lambda_i > 0$ and C a diagonal matrix,

$$C = \begin{pmatrix} c_{11} & 0 \\ 0 & c_{22} \end{pmatrix} \quad \text{with} \quad c_{11} = e^{-\gamma_1 t} \quad \text{and} \quad c_{22} = e^{-\gamma_2 t}$$

Then with $c_i \mathrel{\dot=} c_{ii}$

(2.12)

$$X \cdot C \cdot X^{-1} = \frac{1}{\gamma_2 - \gamma_1} \begin{pmatrix} (\gamma_2 - k_{12}) c_1 + (k_{12} - \gamma_1) \cdot c_2 & k_{21}(c_1 - c_2) \\ k_{12}(c_1 - c_2) & (k_{12} - \gamma_1) c_1 + (\gamma_2 - k_{12}) c_2 \end{pmatrix}$$

with $\gamma_2 - \gamma_1 > 0$, $k_{12} - \gamma_1 \geqq 0$ and $\gamma_2 - k_{12} \geqq 0$ is valid.

Now let the first compartment be the application compartment, such that $d_1 > 0$ and $d_2 = 0$, then

(2.13) $$m(t) = X \cdot C \cdot X^{-1} \cdot d \quad \text{with} \quad d = (d_1, 0)'$$

leads explicitly to

(2.14) $$m_1(t) = \frac{d_1}{\gamma_2 - \gamma_1} \left((\gamma_2 - k_{12}) \cdot e^{-\gamma_1 t} + (k_{12} - \gamma_1) \cdot e^{-\gamma_2 t} \right)$$

$$m_2(t) = \frac{d_1 \cdot k_{12}}{\gamma_2 - \gamma_1} \left(e^{-\gamma_1 t} - e^{-\gamma_2 t} \right)$$

If we determine the bromsulphalein concentration in the blood by means of a sum of exponentials

(2.15) $$y_1(t) = \frac{m_1(t)}{V_1} = a_{11}\, e^{-\gamma_1 t} + a_{21}\, e^{-\gamma_2 t}$$

then the reaction constants k_{ij} and the applied dose d_1 can in this case be calculated except for the distribution volume V_1 of the blood.

From this the curve for the liver, in which measurements cannot be taken, can be simulated.

The expression $y_2(t) = \frac{m_2(t)}{V_1}$ giving the concentration in the liver makes reference to the distribution volume V_1 of the blood.

The following constants and their asymptotic standard deviations were arrived at by means of the maximum likelihood method (fig. 1):

$$a_{11} = 106.0 \pm 10.35 \frac{mg}{ml} \cdot 10^{-3} \; ; \; \gamma_1 = 0.284 \pm 0.037 \frac{1}{min}$$

$$a_{21} = 3.00 \pm 2.51 \frac{mg}{ml} \cdot 10^{-3} \; ; \; \gamma_2 = 0.009 \pm 0.023 \frac{1}{min}$$

and from these the parameters of the model were calculated

$$\frac{d_1}{V_1} = 109.0 \frac{mg}{ml} \cdot 10^{-3} \; ; \; k_{12} = 0.0166 \frac{1}{min}$$

$$k_{21} = 0.122 \frac{1}{min} \; ; \; k_{20} = 0.154 \frac{1}{min}$$

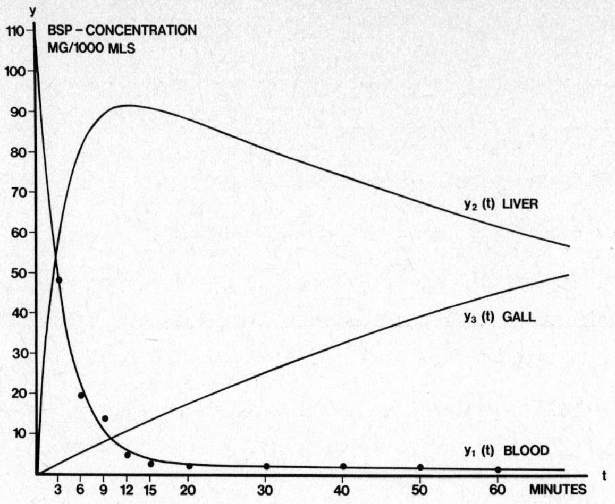

Fig. 1: Clinical Bromsulphalein Test

3. Continuous Application of Medications

We now assume, that a continuous application of medications occurs in the $i^{\underline{th}}$ compartment at constant speed D_i [mass/time]. Such an application form is explainable by infusion or also by the assumption of deep compartments, such as, for instance, fatty tissue, from whose depot there is a constant flow of substance into the blood circulatory system.

The mathematic formulation of a compartment system with deep compartments leads to an inhomogenous initial-value differential equation system

$$(3.1)^* \qquad \dot{m}_i(t) = - \sum_{\substack{j=o \\ j\neq i}}^{n} k_{ij}\, m_i(t) + \sum_{\substack{j=1 \\ j\neq i}}^{n} k_{ji}\, m_j(t) + D_i$$

If we conceive of the application speed D_i as a vector, $D = (D_1, D_2, \ldots, D_n)'$, then this system can be represented in matrix form analogously to (2.1) as

$$(3.1) \qquad \boxed{\dot{m}(t) = K \cdot m(t) + D}$$

The solution of this inhomogenous differential equation system is known to be comprised additively of a special solution of (3.1) and the general solution of (2.1).

Thus

(3.3) $$m(t) = X \cdot e^{\Lambda t} \cdot X^{-1} \cdot m(0) - X(e^{\Lambda t} - I) \cdot \Lambda^{-1} \cdot X^{-1} \cdot D$$

is valid.

Here I is the (n x n) unity matrix and $C = \Lambda^{-1}$ is a diagonal matrix with $c_{ii} = 1/\lambda_i$ for diagonal elements.

(3.3.1) Compartment Models with Deep Compartments

We view the organism as an open system in a flowing balance. A single application of medication disturbs the balance and stimulates a decomposing process, leading back to the condition of equilibrium.

Equilibrium dominates for $\dot{m}_i(t)=0$, thus according to (3.2), it dominates for the case

(3.4) $$\hat{m} = -K^{-1} \cdot D \qquad \text{(equilibrium).}$$

The initial condition for a single application of the dose d_i in a compartment system with deep compartments is therefore

(3.5) $$m(0) = \hat{m} + d$$

When we insert this expression into (3.3) and make allowance for the spectral decomposition (2.6) of $K^{-1} = X \cdot \Lambda^{-1} \cdot X^{-1}$, then

(3.6) $$m(t) = X \cdot e^{\Lambda t} \cdot X^{-1} \cdot d - X \cdot \Lambda^{-1} \cdot X^{-1} \cdot D$$

follows.

2. Example: Glucose Tolerance Test

A simple example for (3.6) is the intravenous glucose tolerance test.

Model 3.1: Clinical Intravenous Glucose Tolerance Test

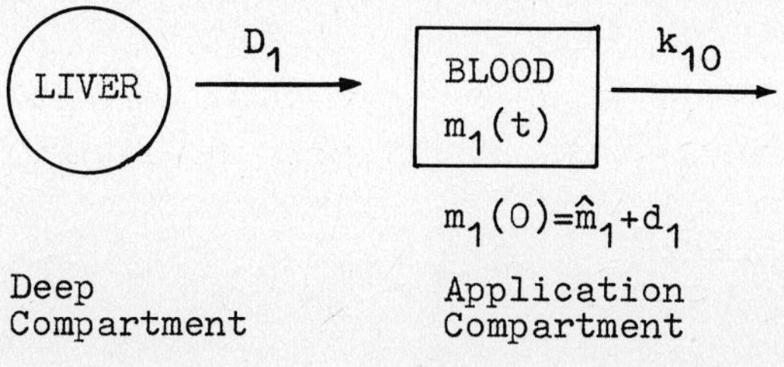

$$\dot{m}_1(t) = -k_{10} \cdot m_1(t) + D_1 \quad \text{with} \quad m_1(0) = \hat{m}_1 + d_1$$

is valid.

For this simple one-dimensional case

$$K = -k_{10}\ , \ X = X^{-1} = 1\ , \ \Lambda = -k_{10} \quad \text{and} \quad \hat{m}_1 = \frac{D_1}{k_{10}}$$

is valid.

According to (3.6) therefore

$$m_1(t) = d_1 \cdot e^{-k_{10} \cdot t} + \frac{D_1}{k_{10}}$$

is valid.

The numerical results arrived at by the inhomogenous exponential trial function

$$y_1(t) = \frac{m_1(t)}{V_1} = a_1 \cdot e^{-k_{10} \cdot t} + a_o$$

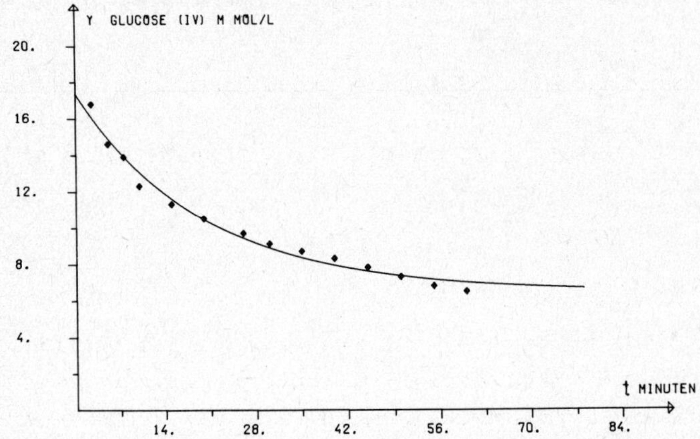

Fig. 2: Intravenous Glucose Tolerance Test

$$a_1 = \frac{d_1}{V_1} = 10.92 \pm 0.43 \ \frac{\text{m mol}}{\text{l}}$$

$$a_o = \frac{D_1}{V_1 \cdot k_{10}} = 6.43 \pm 0.44 \ \frac{\text{m mol}}{\text{l}}$$

$$k_{10} = 0.050 \pm 0.007 \ \frac{1}{\text{min}}$$

(3.6.1) Continuous Infusion

Now we consider the case of continuous infusion in a compartment model without deep compartments. Here D_i [mass/time] means the constant speed of the infusion in the $i^{\underline{th}}$ compartment.

The initial condition is

(3.7) $\quad m(0) = \theta \quad$ (zero vector), and thereby

(3.8) $$\boxed{m(t) = -X\,(e^{\Lambda t} - I) \cdot \Lambda^{-1} \cdot X^{-1} \cdot D}$$

follows from (3.3).

It should be noted that the matrix

$$C = -(e^{\Lambda t} - I) \cdot \Lambda^{-1}$$

is diagonal matrix with elements

(3.9) $$c_{ii} = \frac{1-e^{-\gamma_i \cdot t}}{\gamma_i} \qquad (\gamma_i = -\lambda_i)$$

Let the $k^{\underline{th}}$ compartment be the application compartment, such that $D_k > 0$ and $D_j = 0$ for $i \neq k$, let the $i^{\underline{th}}$ compartment be the measured compartment fitted according to the following exponential trial function:

$$m_i(t) = \sum_{j=1}^{n} a_{ji} \cdot \frac{1-e^{-\gamma_j t}}{\gamma_j}$$

then the relationship between a_{ji} and the parameters k_{ij} is given analogously to (2.10) by

(3.10) $$a_{ji} = D_k \cdot x_{ij} \cdot z_{jk}$$

(3.10.1) Truncated Infusion

It is clear that, by appropriate selection of the initial condition m(0) from (3.3), even general explicit solutions for the case of continuous infusions in systems with deep compartments can be given.

However here we wish only to deal with the case of truncated infusions in systems without deep compartments, a case which often arises in practice.

We assume that in the time interval $0 \leq t < T$ an infusion with constant speed D_k is carried out in the $k^{\underline{th}}$ compartment and is truncated at time t=T.

The law of infusions (3.8) is then valid in the time $0 \leq t < T$. In the time $T \leq t < \infty$ pure elimination follows according to (2.7), thus

$$m(t) = X \cdot e^{\Lambda(t-T)} \cdot X^{-1} \cdot m(T)$$

with the initial condition

$$m(T) = X \cdot (e^{\Lambda T} - I) \cdot \Lambda^{-1} \cdot X^{-1} \cdot D$$

is valid for $t \geq T$.

For truncated infusion in systems without deep compartments

(3.11) $$m(t) = \left\{ \begin{array}{ll} X(e^{\Lambda t}-I) \cdot \Lambda^{-1} \cdot X^{-1} \cdot D & \text{for } 0 \leq t < T \\ X(e^{\Lambda t}-e^{\Lambda(t-T)}) \Lambda^{-1} \cdot X^{-1} \cdot D & \text{for } T \leq t < \infty \end{array} \right\}$$

follows.

3. Example: Truncated Ampicillin Infusion

As in example Nr. 2, we assume a one-compartment model

Model 3.2: Truncated Ampicillin Infusion

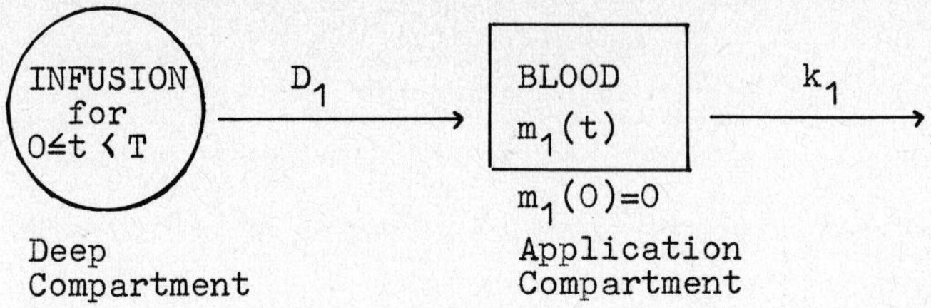

Then

$$m_1(t) = \begin{cases} \dfrac{D_1}{k_1}(1-e^{-k_1 \cdot t}) & \text{for } 0 \leq t < T \\ \dfrac{D_1}{k_1}(e^{-k_1(t-T)} - e^{-k_1 t}) & \text{for } T \leq t < \infty \end{cases}$$

When we calculate these values and their asymptotic standard deviations according to the maximum likelihood method, once with constant length of infusion T = 50 min and then once with simultaneously estimated length of infusion, then figure 3 is valid.

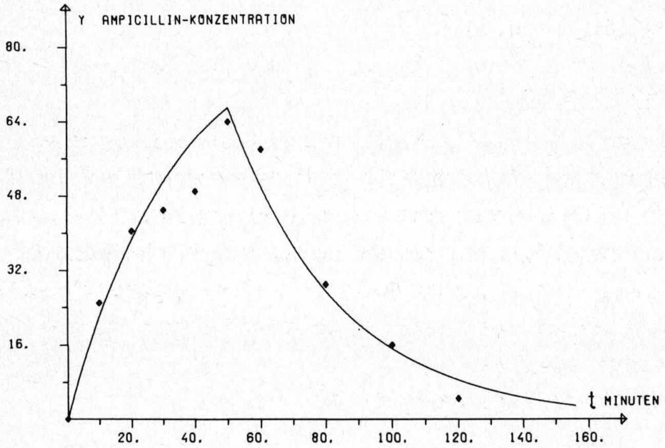

Fig. 3a: Fitted Ampicillin Infusion Curve

with constant length of infusion

$T = 50$, $\dfrac{D_1}{V_1 \cdot k_1} = 86.78 \pm 6.00$, $k_1 = 0.027 \pm 0.0024$

Sum of Squares SSQ = 225.02

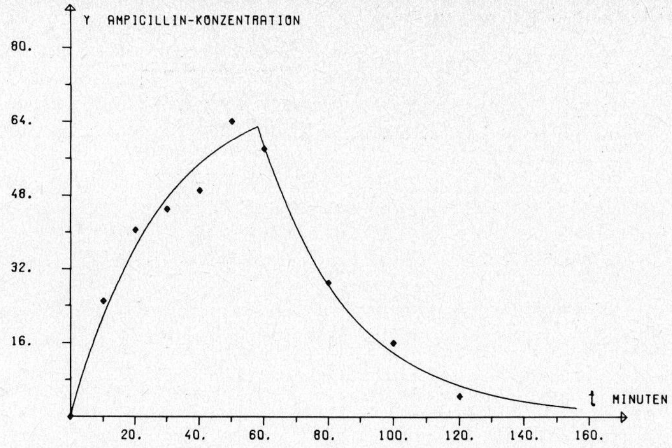

Fig. 3b: Fitted Ampicillin Infusion Curve

with estimated length of infusion

$$T = 57.98 \pm 2.03 \ , \ \frac{D_1}{V_1 \cdot k_1} = 71.81 \pm 4.65 \ , \ k_1 = 0.036 \pm 0.0041$$

Sum of Squares SSQ = 90.17

4. Dost's Law of Area

Dost's law of area (Dost 1968) is often used in practical applications in pharmacokinetics. This principle says that, in a compartment system with a central compartment, the area under the concentration curve of this compartment depends solely upon the central elimination constant and upon the applied dose in a given compartment.

A proof and a generalization of this theorem can be easily derived from the previous chapter.

A summation of the n equations $(3.1)^*$ yields

$$(4.1) \qquad \sum_{i=1}^{n} \dot{m}_i(t) = - \sum_{i=1}^{n} k_{io} \, m_i(t) + \sum_{i=1}^{n} D_i$$

Now, without limiting the generality, let the first compartment be the central one, such that

$$k_{10} > 0 \quad \text{and} \quad k_{io} = 0 \quad \text{for} \quad i=2,3,\ldots,n$$

Then according to the case (3.3.1) of compartment models with deep compartments

$$(4.2)\qquad \sum_{i=1}^{n} \dot{m}_i(t) = -k_{10}\, m_1(t) + \sum_{i=1}^{n} D_i \quad\text{and}\quad m_i(0) = \hat{m}_i + d_i$$

follows.

Now let $\lim_{t\to\infty} m_i(t) = \hat{m}_i$ be the state of equilibrium (3.4) in the $i^{\underline{th}}$ compartment,

$$(4.3)\qquad \boxed{\int_0^\infty (m_1(t)-\hat{m}_1)dt = \frac{\sum_{i=1}^{n} d_i}{k_{10}}} \quad \text{with } \hat{m}_1 = \frac{\sum_{i=1}^{n} D_i}{k_{10}}$$

then follows from the integration of (4.2).

This is a generalization of Dost's principle.

5. Repeated Application of Medication

Another application form often used in medicine is the repeated application of medication. This may be either periodic oral or intramuscular application of medication.

Now we assume that a dose d_i is applied in the $i^{\underline{th}}$ compartment ($i=1,2,\ldots,n$) of a general compartment model without deep compartments at equidistant points in time $j \cdot T$ ($j=0,1,2,\ldots$).

In case at time τ the amount $m(\tau)$ is present in such a compartment system, then

$$(5.1)\qquad m(t) = X \cdot e^{\Lambda(t-\tau)} \cdot X^{-1} \cdot m(\tau) \qquad \text{for} \quad \tau \leq t < \infty$$

is valid according to (2.7).

Now let us consider the time intervals

$$(5.2)\qquad j \cdot T \leq t < (j+1)\cdot T \qquad (j=0,1,2,\ldots)$$

at the beginning of each of which the dose d is applied; it follows from (5.1) by recursion that

$$(5.3)\qquad m(t) = X\cdot e^{\Lambda(t-j\cdot T)}(I-e^{\Lambda(j+1)T})\cdot(I-e^{\Lambda T})^{-1}\cdot X^{-1}\cdot d$$
$$\text{for } j\cdot T \leq t < (j+1)\cdot T$$

This is the general solution for multicompartment models with periodic applications of medications.

It is now possible to represent the solutions, defined only upon intervals, by a solution defined in the complete interval $-T \leq t < \infty$.

Let

$$(5.4)\qquad p(t) = X\cdot(I-e^{\Lambda(t+T)})\cdot(I-e^{\Lambda T})^{-1}\cdot X^{-1}\cdot d$$
$$\text{with} \quad -T \leq t < \infty$$

then

$$(5.5)\qquad \boxed{\begin{array}{ll} m(t) = p(t) - p(t') & \text{for} \quad 0 \leq t < \infty \\ \text{with} \quad t' = t-(j+1)\cdot T & \text{and} \quad -T \leq t' < 0 \end{array}}$$

The course of the curve in each compartment with repeated application of medication can be represented by the superposition of a continuous and differentiable curve $p_i(t)$ and a periodic curve $p_i(t')$.

4. Example: Simulation of Repeated Penicillin Application

Model 5.1: Repeated Penicillin Application

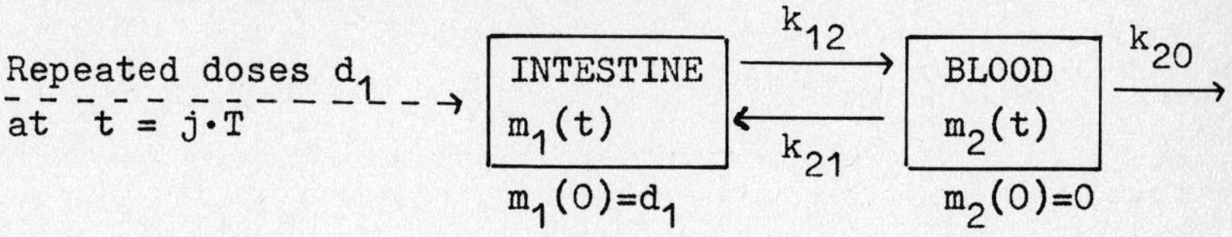

We assume that the dose d_1 is applied in the first compartment at each of the points in time $j \cdot T$ $(j=0,1,2,\ldots)$.
It is now quite simple to calculate the function $p_2(t)$.
We consider the expression (2.12) in example No. 1 and set

$$c_{ii} = \frac{1-e^{-\gamma_i(t+T)}}{1-e^{-\gamma_i \cdot T}}$$

then

$$p(t) = X \cdot C \cdot X^{-1} \cdot d \quad \text{with} \quad d=(d_1,0)'$$

therefore

$$p_2(t) = \frac{d_1 \cdot k_{12}}{\gamma_2 - \gamma_1} \left(\frac{1-e^{-\gamma_1(t+T)}}{1-e^{-\gamma_1 T}} - \frac{1-e^{-\gamma_2(t+T)}}{1-e^{-\gamma_2 T}} \right)$$

is valid for $-T \leq t < \infty$

and

$$\boxed{m_2(t) = p_2(t) - p_2(t')}$$

with $0 \leq t < \infty$ and $t' = t-(j+1) \cdot T$; $-T \leq t' < 0$

Figure 4 shows this approach for the case $k_{12} = 8.7\ h^{-1}$, $k_{20} = 1.37\ h^{-1}$, $k_{21} = 0$ and $d_1 = 1.0\ \frac{\text{unit}}{\text{ml}}$. According to example No. 1 it follows that

$$\gamma_1 = 1.37\ h^{-1} \quad \text{and} \quad \gamma_2 = 8.7\ h^{-1} .$$

The time intervall is set to $T = 0.5$ h.

Taking

$$\bar{p}_2 = \frac{1}{T} \cdot \int_{-T}^{0} p_2(t')dt' = \frac{d_1 \cdot k_{12}}{\gamma_2 - \gamma_1} \left[\frac{1}{1-e^{-\gamma_1 T}} - \frac{1}{1-e^{-\gamma_2 T}} + \frac{1}{\gamma_2 T} - \frac{1}{\gamma_1 T} \right]$$

a mean amount of drug can defined as $m_{2,\mathrm{Mean}}(t) = p_2(t) - \bar{p}_2$.

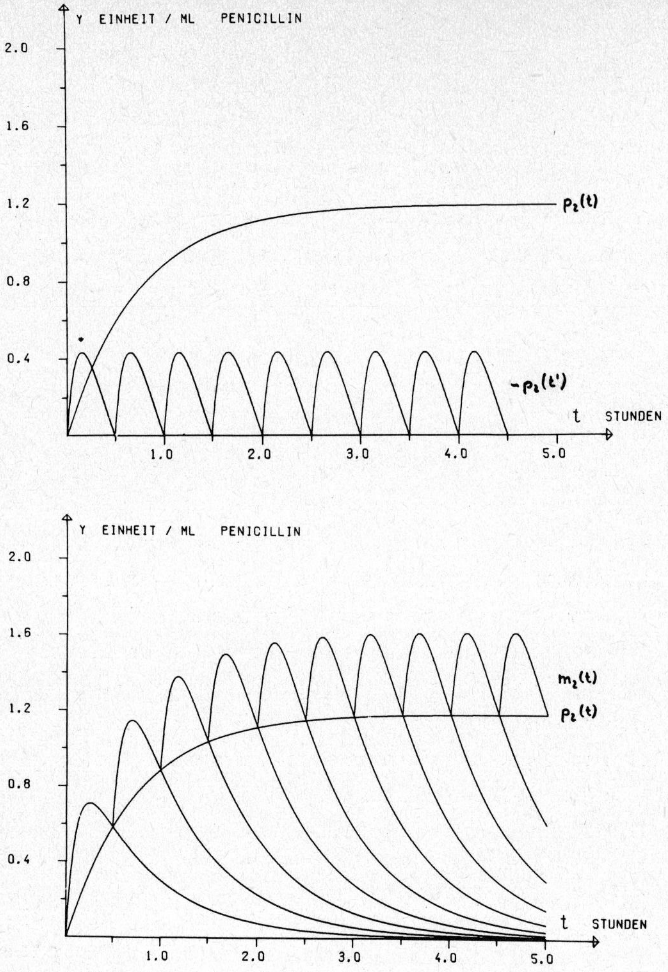

Fig. 4: Repeated Penicillin Application in First Compartment. Superposition Principle Applied to Second Compartment.

Part II: Models for the Pharmacodynamics

1. The basic biological model of pharmacodynamics

As was pointed out in the introductory chapter with "pharmacodynamics" is meant the investigation of the effects of a drug in changing the normal or pathological functions of the organism.
The complicated interaction between drug and drug-response can be schematized by the following compartmental model of Ariëns (1964) and van Rossum (1966), which consists of 4 compartments (see figure 1):

- blood-compartment
- tissue-compartment
- receptor-compartment
- response-compartment.

Between and within these 4 compartments the following changes take place: Into the blood compartment a certain amount D of drug is given (either by a single intravenious injection or during a time interval, e.g. by oral application or infusion). By this a drug concentration C_B in the blood compartment is produced. From the blood compartment the drug is either eliminated or transduced to the tissue compartment. For the drug concentration in the tissue compartment we write C_T. The time behaviour of C_B and C_T can be described by the formulas of pharmacokinetics developed in the first part of this paper.
In the tissue the drug molecules may react with receptors. This process is described within the receptor compartment. As a result of the reaction between drug molecules and receptor a drug receptor complex is built up. The time behaviour of the number of such drug receptor complexes is the interesting function in the receptor compartment. For this number we write Q_R.
By the active drug receptor complexes some stimulus S is given to the organism. The intensity of this stimulus is a function of the amount Q_R of drug receptor complexes. In the lack of better knowledge we may consider this function as a linear one.
The stimulus acts on the last compartment of the scheme, the response compartment. The response E can be a qualitative one (as e.g. death) or a quantitative one (as e.g. change of blood pressure). In the first case, E is a binary function of the intensity of the stimulus S, which takes the two values 1 or 0; in the second case E_A is a graded or continuous function of S.
While pharmacokinetics is mostly concerned with the drug behaviour within and between the blood- and tissue-compartment, the stimulus- and response-behaviour in the receptor- and response-compartment is

investigated by pharmacodynamics. This behaviour depends primary on the amount of drug-receptor complexes. So the pharmacodynamic models are mostly devoted to the interaction between drug molecules and receptors. In the following we shall discuss some deterministic and stochastic models for this interaction. From this interaction mathematical functions for the stimulus and response can be derived by adequate functional assumptions.

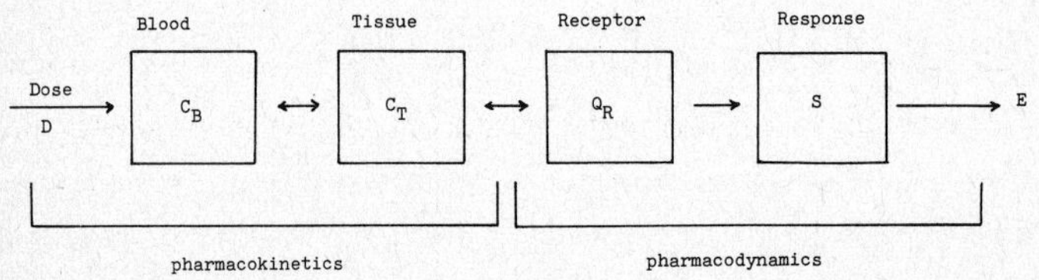

Fig. 1: Scheme of the drug-response interaction in the organism

2. A deterministic model for the interaction between drug and receptors

The interaction between drug and receptors is determined by 2 processes:

a) The association of drug receptor complexes, which can be assumed to be proportional to the drug concentration in the tissue compartment (in the case of one-molecule reaction) or the n-th power of this concentration (in the case of n-molecule reaction) and the number of free receptors.

b) The dissociation of drug receptor complexes, which can be assumed to be proportional to the number of available drug receptor complexes.

From these assumptions we can derive a deterministic mathematical model, if we consider the number of drug receptor complexes Q_R as a continuous function of the time t. The change of this number can be described by the following differential equation:

$$(1) \qquad \frac{dQ_R}{dt} = k_1 \, C_T^n \, (Q - Q_R) - k_2 Q_R$$

together with the initial condition:

$$Q_R = 0 \quad \text{for} \quad t = 0.$$

In this formula there is:

Q_R = number of drug receptor complexes
Q = total number of available receptors
C_T = drug concentration in the tissue
n = number of drug molecules, necessary to build a drug receptor complex
k_1 = association coefficient
k_2 = dissociation coefficient.

The general ideas to this equation were first given by Michaelis and Menten (1913). The equation was later redeveloped by several authors (e.g. by Druckrey (1949), van Rossum (1966)).

If the tissue drug concentration C_T is constant (which can be assumed in the case of drug infusion after sufficient time), the solution of equation (1) is:

$$(2) \qquad Q_R = \frac{k_1 \, C_T^n \, Q}{k_1 \, C_T^n + k_2} \left(1 - e^{-(k_1 C_T^n + k_2)t}\right)$$

The picture of this function is given in figure 2. The number Q_R approaches asymptotically a steady state, which is given by:

$$(3) \qquad Q_{Rst} = \frac{k_1 \, C_T^n \, Q}{k_1 \, C_T^n + k_2}$$

or

$$(4) \qquad \frac{Q_{Rst}}{Q} = g = \frac{1}{1 + \left(\frac{k_2}{k_1}\right)/C_T^n} = \frac{1}{1 + K/C_T^n}$$

g is the ratio of active drug receptor complexes, relative to the total number of receptors.

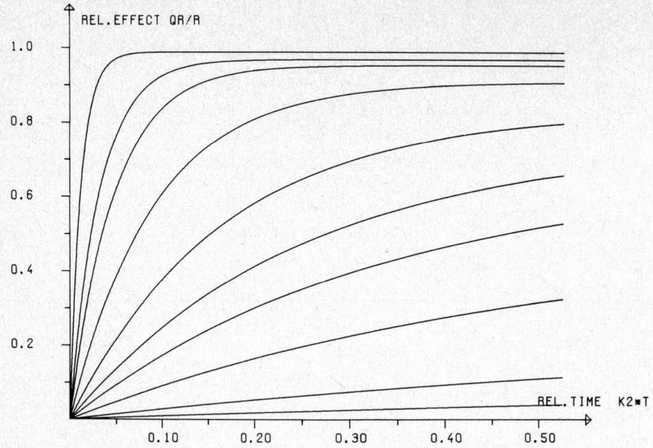

Fig. 2: The relative number of drug receptor complexes as a function of time for different values of drug concentration

So in the steady state the number of drug receptor complexes as a function of the drug concentration in the tissue (which can be considered proportional to the drug dose D) is the well-known logistic function. The constant $K = \frac{k_2}{k_1}$ is called the "Michaelis-Menten" constant or "dissociation-constant". Its reciprocal value $A = \frac{1}{K}$ is called the constant of affinity (see van Rossum (1966)).
If we take the logarithm of the drug concentration C_T, we get:

$$(5) \qquad g = 1 \; / \; \left(1 + 10^{-n(\log C_T - \frac{\log K}{n})}\right)$$

The pictures of the functions (4) and (5) for different time points are shown in figures 3a and 3b. By the parameter K the position on the abscissa and by the parameter n the steepness of the curves is influenced. The value $\frac{\log K}{n}$ corresponds to the logarithm of a "median effective concentration"; that means for $\log C_T = \frac{\log K}{n}$ half of the receptors are occupied by drug molecules.

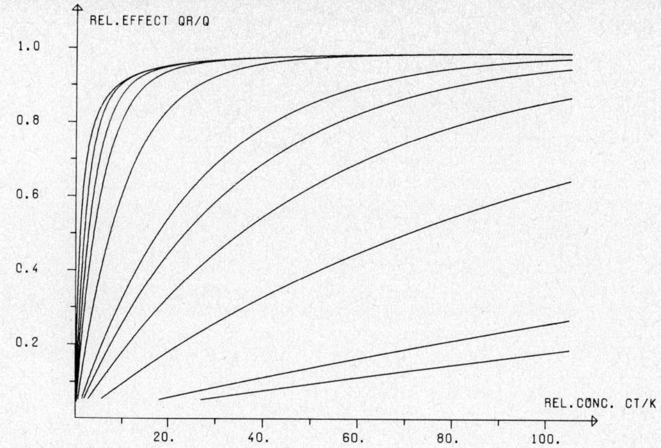

Fig. 3a: The relative effect $g = \frac{Q_R}{Q}$ as a function of the relative concentration $\frac{C_T}{K}$ for different time points

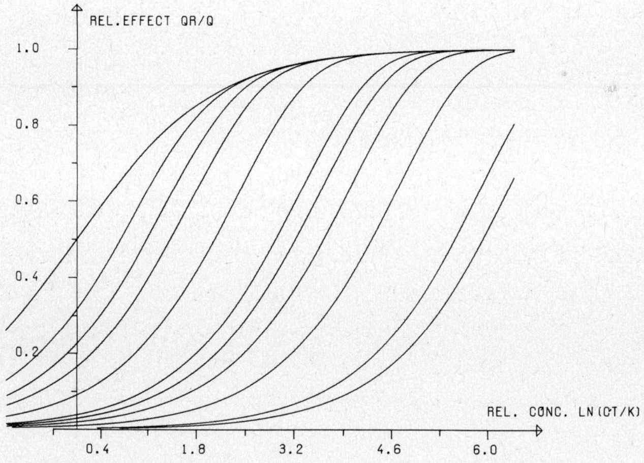

Fig. 3b: The relative effect $g = \frac{Q_R}{Q}$ as a function of the logarithm of the relative concentration $\frac{C_T}{K}$ for different time points

If the drug concentration in the tissue C_T is considered as a function of the time t, the solution of equation (1) looks something more complicated. We get

$$Q_R(t) = Q\left(1-e^{-k_2 t}\cdot e^{-k_1\int_0^t C_T^n(u)du}\left(1+k_2\int_0^t e^{k_2 v}\cdot e^{k_1\int_0^v C_T^n(u)du}dv\right)\right)$$

For the steady state we get the same result as in the case of a constant concentration, where for C_T the steady state value $C_T(\infty)$ must be inserted in the formula. This can be shown either using the equation (1) with $\frac{dQ_R}{dt} = 0$ or the above equation for $t \to \infty$.

3. The relation between the number of drug-receptor complexes and the response

According to the simplified pharmacodynamic model of figure 1 by the drug receptor complexes a stimulus S is produced, which acts on the organism. This stimulus is a mathematical function of the number of drug receptor complexes:

(6) $S = F_S(Q_R) = F_S(g\cdot Q)$

In a first approximation we can assume for F_S a linear function.

(7) $S = \lambda\cdot Q_R = \lambda\cdot Q\cdot g$

The proportionality constant λ is called the "intrinsic activity" (see van Rossum (1966)).

The stimulus S produces in the organism a response E. For this response two different assumptions are important.

a) E is a continuous function of S with values in some real interval (quantitative response, e.g. change in blood pressure).

b) E is a step-function of S, taking only the values 0 (for $S < S_o$) and 1 (for $S \geq S_o$), where S_o is the response threshold (qualitative response, e.g. death).

In the first case one can assume a linear function between E and S, but a logarithmic function has proven as more adequate.

This logarithmic function is the basic assumption in some well known physiological laws, as the Weber-Fechner law or the "initial value law" by Wilder.

In the second case the response threshold S_o is an individual constant, which changes from individual to individual in a random manner. One can assume that these random changes of the response threshold follow a Gaussian or normal distribution. Then the probability of a response for a stimulus S is given by:

(8) $$P = \mathrm{Prob}(S_o \leq S) = \Phi\left(\frac{S-\mu}{\sigma}\right)$$

where μ and σ^2 are the mean and variance of the response levels S_o. For this probability we can write, using formula (4) and (7):

(9) $$P = \Phi\left(\frac{(\sigma Q \cdot g)-\mu}{\sigma}\right) = \Phi\left(\frac{1}{\sigma}\left(\frac{\lambda Q}{1+K/C_T^n} - \mu\right)\right)$$

This formula can be considered as a dose-response relationship, assuming C_T proportional to the given dose D of the drug. In the probit analysis (see Bliss (1934)) a logarithmic normal distribution is assumed as dose-response relationship. Formula (9) shows that the Michaelis-Menten model leads to a somewhat different function representing the normal distribution of a hyperbolic function of the doses as dose-response relationship. One of the intrinsic differences between both functions is the fact, that in formula (9) for $D = kC_T \longrightarrow \infty$ the response probability P approaches an assymptotic value smaller than 1 and for D = 0 it takes a value greater than 0. By experience we would assume that the response probability should approach 1 for sufficient large doses and a value 0 for D = 0. Formula (9) is therefore useful as dose-response function only for medium dose-values.

To get a better approach to realistic dose-response functions, on the basis of normal distributed response thresholds S_o, we must change the assumption of a linear function between stimulus S and ratio of drug receptor complexes g (formula 4). As g varies between 0 and 1, the stimulus S can with the linearity assumption only vary within 0 and a constant value λQ. If we want to get 0 and 1 response probabilities P we must have a possible variation of S between $-\infty$ and $+\infty$.

The most simple function, which fulfills these conditions is:

(10) $$S = \lambda \cdot \log \frac{g}{1-g}$$

With this stimulus function we get as dose-response function:

(11) $$P = \Phi(\beta\,(\log D - \mu))$$

which is exactly the logarithmic normal distribution. The various model parameters are summarized in the 2 parameters β und μ.

4. A simple stochastic model for the interaction between drug and receptors

In reality one cannot expect that the association of drug receptor complexes is a deterministic process; instead we must consider it as a random event. Therefore the number of drug receptor complexes will

be a stochastic time-function rather than a deterministic one.
The first attempt to stochastic considerations for chemical reactions (the drug receptor process can be considered as a special case of chemical reactions) goes back to Kramers (1940) and Delbrück (1940). Later stochastic models were developed by Renyi and Prekopa (1954), Bartholomay (1958), Ishida (1960), McQuarrie (1963) and Tautu (1973). A brief historical review is given in McQuarrie (1967).
In this paper we shall discuss the stochastic equivalent to the Michaelis-Menten model.
Let ξ_t be the number of drug receptor complexes at time t. This should be a random variable dependent on the time t. The set of all such random variables is called a random function or stochastic process.
Let $P_x(t)$ be the probability, that $\xi_t = x$, where x can be any integer number greater than or equal 0. In the time-interval between t and $t + \Delta t$ the following events can take place:

a) The association of a new drug receptor complex; the probability of this event is assumed to be proportional to the drug concentration C_T (we assume here a one-molecule reaction) and the number of free receptors, which is Q-x, if at time t x receptors are occupied. Q is the maximal number of receptors.
b) The dissociation of a drug-receptor complex; the probability of this event is assumed to be proportional to the number of drug receptor complexes at time t.
c) More than 1 drug receptor complex association or dissociation take place; the probability of this event is assumed to be of order $o(\Delta t)$.
d) No change takes place within the time-interval; the probability of this event is the complement of the first 3 probabilities.

The number of drug molecules in the tissue compartment is not considered as random. Instead we assume a definite drug concentration C_T in the tissue compartment. For further simplification we shall assume this concentration as constant during the interesting time. These assumptions may be justified if the drug dose is sufficient large and the tissue compartment has a sufficient low elimination constant or the drug was infused with constant rate.
From these assumptions we get immediately the following difference equation:

$$(12) \qquad P_x(t+\Delta t) = k_1\, C_T(Q-(x-1))P_{x-1}(t)\Delta t + k_2(x+1)\, P_{x+1}(t)\Delta t + (1-k_1\, C_T(Q-x) - k_2x)P_x(t)\Delta t + o(\Delta t)$$

Here k_1 is the stochastic association rate and k_2 the stochastic dissociation rate.

Taking the limit $\Delta t \to 0$ we get from this equation a system of difference differential equations for the probabilities $P_x(t)$:

$$(13) \qquad \frac{dP_x(t)}{dt} = k_1C_T(Q-(x-1))P_{x-1}(t) + k_2(x+1)P_{x+1}(t) - (k_1C_T(Q-x) + k_2x)\, P_x(t)$$

As initial condition we have:

$$(14) \qquad P_x(0) = \begin{cases} 1 & \text{for} \quad x = 0 \\ 0 & \text{for} \quad x > 0 \end{cases}$$

This system of equations can be transformed in a partial differential equation for the generating function:

$$F(s,t) = \sum_{x=0}^{G} s^x P_x(t)$$

From equation (13) we get:

$$(15) \qquad \frac{\partial F(s,t)}{\partial t} = (k_2 + (k_1C_T - k_2)s - k_1C_Ts^2)\,\frac{\partial F(s,t)}{\partial s} + k_1C_TQ\,(s-1)\,F(s,t)$$

and the initial condition: $F(s,0) = 1$.

From this equation all properties of the probability distributions $P_x(t)$ can be derived:

For the mean: $m(\xi(t)) = E(\xi_t) = \frac{\partial F(1,t)}{\partial s}$ we get from (15):

$$(16) \qquad \frac{dm}{dt} = -(k_1\, C_T + k_2)m + k_1C_TQ$$

and $m(\xi(0)) = 0$.

This equation is identical with the differential equation (1) for the deterministic model (for n=1). Therefore the mean $m(\xi(t))$ of the simple stochastic model is identical with the time-function of the deterministic model:

$$(17) \qquad m(\xi(t)) = \frac{k_1\, C_T\, Q}{k_1C_T+k_2}\,(1 - e^{-(k_1C_T+k_2)t})$$

For the variance of $\xi(t)$ we have the general valid relation:

$$(18) \qquad var(\xi(t)) = \frac{\partial^2 F(1,t)}{\partial s^2} + m(t) - m(t)^2$$

Therewith we get from equation (15)

$$(19)\qquad \frac{d\,var(\xi(t))}{dt} = -2(k_1C_T+k_2)var(\xi(t))-(k_1C_T-k_2)m(t) + k_1C_TQ$$

This differential equation has for the initial condition $var(\xi(0))=0$ the solution:

$$(20)\qquad var(\xi(t)) = \frac{Q\,k_1\,C_T}{(k_1C_T+k_2)^2}\Big[k_2(1-e^{-2(k_1C_T+k_2)t}) + (k_1C_T-k_2)(1-e^{-(k_1C_T+k_2)t})e^{-(k_1C_T+k_2)t}\Big]$$

So the mean and the variance of $\xi(t)$ are both increasing with the total number Q of receptors. But if we take the relative number of occupied receptors: $g(t) = \frac{\xi(t)}{Q}$, we see from equations (10) and (12), that the mean of g(t) is independent of Q and the variance is decreasing with Q (as we would expect from the binomial distribution). So for sufficient great Q we can neglect random fluctuations and consider g(t) well represented by the deterministic model.

In the steady state case we have $\frac{dF}{dt} = 0$. So the generating function $F_{st}(s)$ in the steady state follows the differential equation:

$$(21)\qquad (k_2+(k_1C_T-k_2)s-k_1C_Ts^2)\,\frac{dF_{st}}{dt} = -k_1C_TQ(s-1)F_{st}$$

The solution of this equation under the condition $F_s(1) = 1$ is:

$$(22)\qquad F_{st}(s) = \Big(\frac{k_1C_T}{k_1C_T+k_2}\cdot s + \frac{k_2}{k_1C_T+k_2}\Big)^Q$$

This is the generating function of a binomial distribution with the probability $p = \frac{k_1\,C_T}{k_1C_T+k_2}$ and the total number n = G.

Therefore we get immediately for the limiting distribution:

$$(23)\qquad P(\xi_{st} = x) = \binom{Q}{x}\frac{(k_1C_T)^x\,k_2^{(Q-x)}}{(k_1C_T+k_2)^Q}$$

$$m(\xi_{st}) = Q\cdot\frac{k_1C_T}{k_1C_T+k_2}$$

$$var(\xi_{st}) = Q\cdot\frac{k_1k_2C_T}{(k_1C_T+k_2)^2}$$

This agrees with the limits of equation (17) and (20) for $t\to\infty$, when the steady state is reached. If we assume, that for the association of a drug receptor complex $n > 1$ drug molecules are necessary, in all the foregoing formulas - as in the deterministic model - the concentration C_T must be replaced by the n-th power C_T^n. This is but only valid, if C_T is deterministic and not influenced by the process of complex association. As was pointed out, this assumption is agreeable, if the concentration C_T is sufficient high, so that the number of drug molecules in the tissue compartment is very large compared with the number G of possible receptors.

5. Models with a stochastic behaviour of the drug molecules in the tissue compartment

One of the most important extensions of the foregoing discussed stochastic model concerns the behaviour of the drug molecules in the tissue compartment. The assumption of a deterministic behaviour of the drug molecules is not further valid, if the molecule number in the tissue compartment is not too large, so that it is influenced by the association or dissociation of drug receptor complexes. As both processes are stochastic, the number of drug molecules in the tissue compartment must also be considered as a random number. Such models are discussed in papers by Bartholomay (1964) and Tautu (1973). In these papers there are given further references. We shall summarize in this chapter the most important features of these models as compared with the simple stochastic model with constant molecule numbers:

The behaviour in the tissue compartment may be described now by 3 random numbers:

ξ_1 = number of drug molecules

ξ_2 = number of free receptors

ξ_3 = number of drug receptors.

The actual values of these random numbers are called n_1, n_2 and n_3. These are integers among which the following conditions should hold:

condition a) $\quad n_2 = G - n_3$

condition b) $\quad n_1 = D - n_3$

Here G is the maximal number of receptors and D the maximal number of drug molecules in the compartment. The conditions a) and b) mean, that the system is closed according to both these numbers. We further assume, that $0 \leq n_2 \leq G < D$ and $\xi_3 = 0$ with probability 1 for t = 0.

As in the foregoing chapter the process should be characterized by the following transition probabilities:

- the association probability of a drug response complex in time $(t,t+\Delta t)$, which is $k_1 n_1 \cdot n_2 \Delta t$, if at time t there are n_1 drug molecules and n_2 free receptors in the compartment; k_1 is the stochastic association rate.
- the dissociation probability of a drug receptor complex in time $(t,t+\Delta t)$, which is $k_2 n_2 \Delta t$, if at time t there are n_3 drug receptor complexes; k_2 is the stochastic dissociation rate.

- the probability of more than one transition is of order $o(\Delta t)$.
Then we have for the probability $P(n_1,n_2,n_3;\, t+\Delta t)$, that at $t+\Delta t$ there are n_1 molecules, n_2 free receptors and n_3 drug receptor complexes in the compartment:

$$(24)\quad \begin{aligned} P(n_1,n_2,n_3;t+\Delta t) = {} & k_1(n_1+1)(n_2+1)P(n_1+1,n_2+1,n_3-1;t)\Delta t \\ & + k_2(n_3+1)P(n_1-1,n_2-1,n_3+1;t)\Delta t \\ & + (1-(k_1n_1n_2+k_2n_3))P(n_1,n_2,n_3;t)\Delta t \\ & + o(\Delta t) \end{aligned}$$

For this we get the system of difference differential equations:

$$(25)\quad \begin{aligned} \frac{dP(n_1,n_2,n_3;t)}{dt} = {} & k_1(n_1+1)(n_2+1)P(n_1+1,n_2+1,n_3-1;t) \\ & + k_2(n_3+1)P(n_1-1,n_2-1)n_3+1;t) \\ & - (k_1n_1n_2+k_2n_3)p(n_1,n_2,n_3;t) \end{aligned}$$

The numbers n_1,n_2,n_3 are bound by the conditions a) and b). Therefore 2 of them can be replaced by the third one and we can eliminate the two from equation (25). If we substitute for n_1 and n_2 the variable n_3 according to the conditions a) and b), we get:

$$(26)\quad \begin{aligned} \frac{dP(n_3,t)}{dt} = {} & k_1(DG-(D+G)(n_3-1)+(n_3-1)^2)P(n_3-1;t) \\ & + k_2(n_3+1)P(n_3+1;t) \\ & - (k_1(DG-(D+G)n_3+n_3^2)+k_2n_3)P(n_3;t) \end{aligned}$$

From this equation we derive for the generating function

$$F(s,t) = \sum_{n_3} s^{n_3}P(n_3;t):$$

the following partial differential equation:

$$(27)\quad \frac{\partial F(s,t)}{\partial t} = k_1s^2(s-1)\,\frac{\partial^2F(s,t)}{\partial s^2} - (k_1s(s-1)\,(D+G-1)+k_2(s-1)\,\frac{\partial F(s,t)}{\partial s} + k_1DG(s-1)F(s,t).$$

Comparing this equation with equation (15) the most remarkable difference is the new existence of partial derivations of second order in equation (27). This is due to the existence of quadratic terms in n_3, which appear in equation (26). By these terms equations (26) and (27) cannot be solved explicitly. But there exist some approximate solution approaches, which are summarized and discussed in McQuarrie (1967) and Herrmann (1976).
From equation (27) we derive a differential equation, for $m(\xi_3(t))$ the expected value of $\xi_3(t)$ using the relations:

$$E(\xi_3(t)) = \left.\frac{\partial F(s,t)}{\partial s}\right|_{s=1} \quad\text{and}\quad E(\xi_3^2(t)) = \left.\frac{\partial^2F(s,t)}{\partial s^2}\right|_{s=1} - E(\xi_3(t)):$$

(28) $$\frac{dm(\xi_3(t))}{dt} = k_1 E(\xi_3^2(t)) - (k_1(D+G)+k_2)m(\xi_3(t)) + k_1 DG$$

This equation can only be solved, if the second moment $E(\xi_3^2(t))$ is known. This is in general not the case.

The same situation exists also for higher moments. In each of these equations for the moment of order n there is a term of order n+1. So by this procedure we cannot get explicit expressions for the moments of the distribution of $\xi_3(t)$.

One of the most important things is the fact, that equation (28) differs from the corresponding differential equation of the deterministic model. Therefore the mean of the stochastic model differs from the solution of the deterministic model. The deterministic model is characterized by the equation:

(29) $$\begin{aligned}\frac{dQ_R}{dt} &= k_1(D-Q_R(t))\,(G-Q_R(t)) - k_2 Q_R \\ &= k_1 Q_R^2(t) - (k_1(D+G)+k_2)Q_R(t) + k_1 DG\end{aligned}$$

Here $Q_R(t)$ is the deterministic number of drug receptor complexes as a function of time.

Both equations differ in the first term, which is in the stochastic equation the second moment and in the deterministic equation the square of the number of drug receptor complexes. As we have

$$E(\xi_3^2(t)) = var(\xi_3(t)) + m^2(\xi_3(t)) < m^2(\xi_3(t))$$

the mean of the stochastic model is always less than the corresponding value of the deterministic model. This remains valid also for the limit case of steady state for $t \to \infty$.

A similar situation arises, if for a drug receptor complex more than one molecule is necessary and the number of molecules in the tissue compartment depends on the number of drug receptor complexes according to condition b). Then the association probability is proportional to a powerfunction of the number n_3 of drug receptor complexes of order n+1, if n molecules are necessary for a complex association.

A detailed discussion of the bimolecular reaction process is given in McQuarrie (1967). The reader is referred to this publication for further information.

6. Some general approaches to stochastic models for the drug-receptor interaction

The models discussed in the last chapters belong to the class of so-called "birth and death"-processes. The association of drug receptor complexes is considered as a "birth" event, the dissociation as a

"death". In the simple model of chapter 4, the "birth" probability is proportional to the existing number of complexes; in the extended model this probability is proportional to a quadratic function of the complexes. Both processes are well known in the mathematical population theory; the first one as linear birth process or Malthus-process, the second one as logistic process or Verhulst-process.
It should be mentioned here, that quite different ideas and model approaches can be developed, which will result in similar mathematical formulas for the drug receptor process, but give new insight to the quantitative behaviour of the process.
One of the most important of these approaches is the "General Catalytic Queue Process" by Bartholomay (1964). In this approach the Theory of Queues is used as mathematical tool for the description of the drug receptor interaction. The drug molecules in the tissue compartment (biophase) are considered as "customers" each waiting to have a "service" performed by one of the receptors, which is considered as "server". The number of customers served by a given server at the same time may be either a fixed constant or a random number. The "service times" are values of a continuous random variable $\tau > 0$ with distribution function $G(t) = \text{Prob}(\tau \leq t)$. According to the Michaelis-Menten model, 3 possible outcomes of a serving event are possible:

- the drug receptor complex association
- the complex dissociation
- both - drug molecule and receptor - finish the event without result.

The process is then analogous to the "age-dependent birth and death process" treated by Bellmann and Harris (see Harris (1963)). For the generating function F(s,t) of this process the following functional equation is valid:

$$F(s,t) = s(1-G(t)) + \int_0^t h(F(s,t-u))dG(u)$$

where G(t) is the distribution function of the service times and h(s) is the generating function of the outcomes of the serving event.

This equation may be used to give some approximative results for the mean and asymptotic behaviour of the process.
Another approach comes from the "Occupancy Theory". This theory was first developed in connection with the kinetic gas theory by Boltzmann and later extended by Bose-Einstein and Fermi-Dirac (see Feller (1957)). The basic model of this theory considers a number of baskets, in which balls are thrown at random. The interesting event is the occupancy of one of the baskets by a ball.

To use this theory for pharmacodynamics the baskets must be interpreted as receptors and the balls as drug molecules. The occupancy event is the association of a drug receptor complex. If we assume, that the molecules arrive at the receptors at random time τ with exponential distribution: $\mathrm{Prob}(\tau \leq t) = 1-e^{-\lambda t}$, that at each arrival a drug receptor complex is formed and no dissociation takes place, the number of drug receptor complexes at time t has a binomial distribution with the mean:

$$Q_R = Q(1 - e^{-\lambda t})$$

where Q is the total number of receptors. This corresponds with equation (2), if we assume the mean number of arrivals per time unit proportional to the drug concentration C_T: $\lambda = k_1 C_T$ and a total lack of dissociation ($k_2 = 0$).

The existence of dissociation can be included in the model by considering the dissociation as a stochastic point process with an exponential distribution of the times τ_2 of events: $\mathrm{Prob}(\tau_2 \leq t) = 1-e^{-k_2 t}$ and independent of the association event. If both events are combined, we get as formula for the probability of the existence of a drug receptor complex at time t:

$$\mathrm{Prob} = 1 - e^{-(k_1 C_T + k_2)t}$$

This corresponds again with the Michaelis-Menten equation (2). These considerations demonstrate, that the occupancy theoretical approach is a feasible one for pharmacodynamic models. It can be used, to extend the Michaelis-Menten model to more complicate situations. Such situations may arise if

- the distribution of arrival times τ of molecules and dissociation times τ_2 of complexes are general distribution functions.
- for the drug receptor complex association more complicated conditions, than the simple arrival of a molecule to a receptor must be fulfilled.

The later situation was extensively investigated by Herrmann (1976a), who derived formulas for the association probabilities under rather general association conditions, including the case of different types of molecules and/or receptors, which play some role in stereochemical analysis.

7. The secondary pharmacological process as model for stimulus and response

If the process of drug receptor complex association is considered as stochastic, stimulus and response to this association will also be stochastic events, even if they are assumed as deterministic functions of the number of complexes. But it is more appropriate to assume stimulus and response as stochastic events depending in a random manner from the stochastic process of complex association. Stochastic models for these processes were introduced by Tautu (1973) and called "secondary pharmacological process".
We will briefly sketch the main ideas of these processes: Let τ_1, $\tau_2, \dots, \tau_k, \dots$ be the time-points of a drug receptor complex association. These are random variables, whose compound distribution can be derived by the process discussed in the foregoing chapters. Each complex association has as result an increase of stimulus and response intensity, which can be considered as a deterministic or stochastic function $h(s,\alpha)$, where s is the time counted from the time of the complex association as initial event and α is a stochastic parameter. It is assumed that $h(s,\alpha)$ tends to 0 with probability 1 if $s \longrightarrow \infty$.

The stimulus or response at time t is considered as the sum of all intensity increase achieved till the time t by complex associations. That means:

(30) $$\eta(t) = \sum_{0 \leq \tau_k \leq t} k(t-\tau_k, \alpha_k)$$

$\eta(t)$ is the total intensity of the stimulus or response at time t. This is called the "secondary pharmacological process". Formula (30) can be formulated more elegantly, if the point process $\zeta(t)$ is introduced, which takes the value 1 for t_k if at this time a complex association occurs,and the value 0 for all other times t. Then the secondary process $\eta(t)$ can be formulated:

(31) $$\eta(t) = \zeta(t) * h(t,\alpha)$$

where * is the symbol of the convolution of the two stochastic processes.
Some examples of the application of secondary processes to stochastic response functions are given by Tautu (1973). The reader is referred to this literature for further discussion.

8. Some concluding remarks

It was the main intention of this paper, to give a broad review over the most important different approaches to mathematical models for pharmacokinetic and pharmacodynamic processes. In pharmacokinetic the basic idea is the compartmental model, in pharmacodynamics the Michaelis-Menten model. Both models can be formulated as deterministic or stochastic ones. If the number of molecules in the compartments are sufficient large, the stochastic formulation converges to the deterministic one. So for most of the practical applications the deterministic models can be considered as adequate and correspond to the real situation rather well. The stochastic models give some additional insight to the behaviour of the random fluctuations and to the possible tendencies in more complex situations. One of the disadvantages of the stochastic models is the lack of explicit formulas, even for the first and second moments in more complex situations. Here only by approximation methods or by direct simulation with Monte-Carlo-methods numerical results can be obtained. This reduces the practical applicability of the more complex stochastic models. But in spite of this fact the stochastic formulation of biological processes is very valuable to get a more detailed insight in the quantitative behaviour of biological processes and is therefore an indispensable intellectual tool of theoretical biology.

References

ARIENS, E.J. (ed.): Molecular Pharmacology. Vol. I. New York 1964.

BARTHOLOMAY, A.F.: Stochastic models for chemical reactions. I.: Theory of the unimolecular reaction process. Bull. Math. Biophys. 20, 175-190 (1958).

BARTHOLOMAY, A.F.: Stochastic models for chemical reactions. II: The unimolecular rate constant. Bull. Math. Biophys. 21, 363-373 (1959).

BARTHOLOMAY, A.F.: The general catalytic queue process. In: Stochastic Models in Medicine and Biology (ed. J. GURLAND). Madison 1964.

BERMAN,M., and SCHOENFELD, R.: Invariants in experimental data and linear kinetics and the formulation of models. Journ. Appl. Physics 27, No. 11, 1161-1370 (1956).

BLISS, C.J.: The method of probits. Science 79, 38-39 and 409-410 (1934).

DELBRÜCK, M.: Statistical fluctuations in autocalatytic reactions. Journ. of Chem. Phys. 8, 120-124 (1940).

DOST, F.H.: Grundlagen der Pharmakokinetic. Stuttgart:G. Thieme 1968.

DRUCKREY, H., and KUPFMULLER,K.: Dosis und Wirkung. Aulendorf 1934.

FELLER, W.: An introduction to probability theory and its application. 2nd ed. New York: J. Wiley 1957.

GLADTKE, E., and HATTINGBERG, H.M.: Pharmakokinetik. Berlin-Heidelberg, New York: Springer 1973.

HARRIS, T.E.: The theory of branching processes. Berlin-Göttingen-Heidelberg: Springer 1963.

HERRMANN, H.: Stochastische Modelle zum Reiz-Antwort-Prozeß der Pharmakon-Receptor-Theorie. Thesis, Abteilung für Biometrie der Medizinischen Hochschule Hannover, Hannover (1976).

HERRMANN, H.: Treffer- und Besetzungstheorie und ihre Anwendung in der Biologie. (In preparation) (1976a).

ISHIDA, K.: The probability approach in absorption processes. Nippon Kagadu Zasshi 81, 524-529 (1960).

JACQUEZ, J.A.: Compartmental analysis in biology and medicine. Amsterdam: Elsevier 1973.

KRAMERS, H.A.: Brownian motion in a field of force and the diffusion model of chemical reactions. Physica 7, 284-304 (1940).

MCQUARRIE, D.A.: Stochastic approach to chemical kinetics. Journ. Appl. Prob. 4, 413-478 (1967).

MICHAELIS, L., and MENTEN, M.: Biochemische Zeitschrift 49, 333 (1913).

PHARMA-KODEX. Bundesverband der Pharmazeutischen Industrie e.V., Frankfurt/Main.

PREKOPA, A.: Statistical treatment of the degradation process of long chain polymers. Magyar Tud. Akad. Alkalm. Mat. Int. Közl. 2, 103-123 (1954).

RENYI, A.: Treatment of chemical reactions by means of the theory of stochastic processes. Magyar Tud. Akad. Alkalm. Mat. Int. Közl. 2, 93-101 (1954).

RESCIGNO, A., and SEGRE, G.: Drug and Tracer Kinetics. Massachusetts: Blaisdell 1966.

ROSSUM van, J.M.: Die Pharmakon-Rezeptor-Theorie als Grundlage der Wirkung von Arzneimitteln. Arzneimittel-Forschung 16, 1412-1426 (1966).

SHEPPARD, C.W., and HOUSEHOLDER, A.S.: The mathematical basis of the interpretation of tracer experiments in closed steady-state systems. Appl. Physics 22, No. 4, 510-520 (1951).

TAUTU, P.: Probability models in pharmacodynamics; a prospect. Mimeographed paper by Deutsches Krebsforschungszentrum, Heidelberg (1973).

D i s c u s s i o n

Concerning the paper of Richter

Nijssen:
Which programs did you use for your simulations ?

Richter:
The subroutine HPGG (IBM sientific subroutine package) was used which is based on Hamming's predictor-corrector method.

Concerning the paper of Müller-Schauenburg

Steinijanso:
I am afraid that the problem of metabolism was neglected in your lecture. In particular, if a substantial metabolic transformation takes place during the first liver passage the conclusions drawn from intravenous study alone may be totally misleading for the oral route of drug administration.

Müller-Schauenburg:
Metabolism is mainly a type of excretion from a lateral compartment. Some of my applications admit these processes, as stated in the introduction. The very first-pass effect e.g. an invasion into the system via a lateral compartment /22/ has not been considered here.

Pallaske:
Time constants γ_i may only be extracted from measured data curves by exponetial analysis. Analysing more than two superimposed exponential terms is only possible if very exactly measured data are available.

Müller-Schauenburg:
I agree.

Concerning the paper of Feldmann-Schneider

Neiss:
If the K_{ij} in your model are known then your method is a proper tool to describe the m (t) in the compartments. Treating practical problems one is often interested to estimate k_{ij}. Therefore you need the functional form of the m (t) to fit data.
Here you can give in a closed form the relation between the parameters of the curve and the k_{ij}.

Feldmann:
The parameters k_{ij} are indeed the ones which one wants to estimate for practical approximation problems since they may be interpreted biologically as elimination or invasion constants. The purpose of the method under discussion is to solve the differential equations explicitly posed by the pharmacokinetik model and to calculate the m (t) as exponential summations usually used for fitting. The relationship of constants a_{ij} and γ_j of the exponential summations and the parameters k_{ij} of the differential equations is stated in the eigenvalue problem of my paper. There is a non-linear relationship between the parameters and the constants as outlined in part I, section 2.

Pallaske:
Is it possible to find k_{ij} if the measurement can only be done in one compartment ?

Feldmann:
Using a general n compartment model without deep compartments there may be n x (n-1) parameters k_{ij} and n initial conditions d_i.
If the measured data are from only one compartment only 2 n constants may be estimated.
Should there be no more than 2 n parameters k_{ij} in a model it may be possible, in special cases like the bromsulphalein test, to extrapolate from measurements on one compartment the behavior of the total system.

Editors: K. Krickeberg;
S. Levin; R. C. Lewontin;
J. Neyman; M. Schreiber

Biomathematics

Vol. 1: **Mathematical Topics in Population Genetics**
Edited by K. Kojima
55 figures. IX, 400 pages. 1970
ISBN 3-540-05054-X

This book is unique in bringing together in one volume many, if not most, of the mathematical theories of population genetics presented in the past which are still valid and some of the current mathematical investigations.

Vol. 2: E. Batschelet
Introduction to Mathematics for Life Scientists
200 figures. XIV, 495 pages. 1971
ISBN 3-540-05522-3

This book introduces the student of biology and medicine to such topics as sets, real and complex numbers, elementary functions, differential and integral calculus, differential equations, probability, matrices and vectors.

M. Iosifescu; P. Tautu
Stochastic Processes and Applications in Biology and Medicine

Vol. 3: Part 1: Theory
331 pages. 1973
ISBN 3-540-06270-X

Vol. 4: Part 2: Models
337 pages. 1973
ISBN 3-540-06271-8

Distribution Rights for the Socialist Countries: Romlibri, Bucharest

This two-volume treatise is intended as an introduction for mathematicians and biologists with a mathematical background to the study of stochastic processes and their applications in medicine and biology. It is both a textbook and a survey of the most recent developments in this field.

Vol. 5: A. Jacquard
The Genetic Structure of Populations
Translated by B. Charlesworth; D. Charlesworth
92 figures. Approx. 580 pages. 1974
ISBN 3-540-06329-3

Population genetics involves the application of genetic information to the problems of evolution. Since genetics models based on probability theory are not too remote from reality, the results of such modeling are relatively reliable and can make important contributions to research. This textbook was first published in French; the English edition has been revised with respect to its scientific content and instructional method.

Springer-Verlag
Berlin
Heidelberg
New York